Earth Sciences in the 21st Century

AF587942

Earth Sciences in the 21st Century

Neogene Deep Water Benthic Foraminifera from the Indian Ocean – A Monograph
Anil Kumar Gupta (Editor)
2022. ISBN: 979-8-88697-385-3 (Hardcover)
2022. ISBN: 979-8-88697-412-6 (eBook)

Observing Micrometeorology: A Personal Tour through an Evolving Science
Bruce B. Hicks (Editor)
2022. ISBN: 979-8-88697-369-3 (Hardcover)
2022. ISBN: 979-8-88697-426-3 (eBook)

The Basics of Groundwater
Melissa Clutter, PhD (Author)
Chloé Fandel, PhD (Author)
Ty Ferré, PhD (Author)
2022. ISBN: 978-1-68507-740-2 (Hardcover)
2022. ISBN: 978-1-68507-874-4 (eBook)

Geo-Information Technology in Earth Resources Monitoring and Management
Dr. Varun Narayan Mishra (Editor)
Dr. Praveen Kumar Rai (Editor)
Dr. Prafull Singh (Editor)
2021. ISBN: 978-1-53619-669-6 (Hardcover)
2021. ISBN: 978-1-53619-796-9 (eBook)

More information about this series can be found at
https://novapublishers.com/product-category/series/earth-sciences-in-the-21st-century/

Veronica J. Avila
Editor

Transition Metals

An Overview

Copyright © 2023 by Nova Science Publishers, Inc.

All rights reserved. No part of this book may be reproduced, stored in a retrieval system or transmitted in any form or by any means: electronic, electrostatic, magnetic, tape, mechanical photocopying, recording or otherwise without the written permission of the Publisher.

We have partnered with Copyright Clearance Center to make it easy for you to obtain permissions to reuse content from this publication. Please visit copyright.com and search by Title, ISBN, or ISSN.

For further questions about using the service on copyright.com, please contact:

Copyright Clearance Center
Phone: +1-(978) 750-8400 Fax: +1-(978) 750-4470 E-mail: info@copyright.com

NOTICE TO THE READER

The Publisher has taken reasonable care in the preparation of this book but makes no expressed or implied warranty of any kind and assumes no responsibility for any errors or omissions. No liability is assumed for incidental or consequential damages in connection with or arising out of information contained in this book. The Publisher shall not be liable for any special, consequential, or exemplary damages resulting, in whole or in part, from the readers' use of, or reliance upon, this material. Any parts of this book based on government reports are so indicated and copyright is claimed for those parts to the extent applicable to compilations of such works.

Independent verification should be sought for any data, advice or recommendations contained in this book. In addition, no responsibility is assumed by the Publisher for any injury and/or damage to persons or property arising from any methods, products, instructions, ideas or otherwise contained in this publication.

This publication is designed to provide accurate and authoritative information with regards to the subject matter covered herein. It is sold with the clear understanding that the Publisher is not engaged in rendering legal or any other professional services. If legal or any other expert assistance is required, the services of a competent person should be sought. FROM A DECLARATION OF PARTICIPANTS JOINTLY ADOPTED BY A COMMITTEE OF THE AMERICAN BAR ASSOCIATION AND A COMMITTEE OF PUBLISHERS.

Library of Congress Cataloging-in-Publication Data

ISBN: 979-8-88697-646-5

Published by Nova Science Publishers, Inc. † New York

Contents

Preface

This book contains five selected chapters which provide an overview of transition metals. Chapter One explores transition metal based cathode materials for lithium-ion cells. Chapter Two reviews recent advances on the functionalization of 4-quinolone derivatives via transition metal catalysis. Chapter Three examines the screened exchange interaction in bulk at the interface of transition metals. Chapter Four summarizes the synthesis, characterization, and use of the copper(II)-1,4-bis-(3-hexafluoroacetylacetonate)benzene framework. Chapter Five reviews iron(III) nitrate as a catalyst in acid-catalyzed reactions.

Chapter 1 - Lithium-ion batteries (LIBs) are considered to be the most promising power source for electric vehicles (EVs) and portable electronic devices. Transition metals are the cornerstone of cathode materials, the key component that determine the overall performance of LIBs. This chapter reviews the advances in exploring and developing various transition metal oxides as high-performance cathode materials for LIBs. The primary focus is on the structure and especially the electrochemical properties of several classes of transition metal-based cathodes. Moreover, the strategies capable of overcoming the current limitations of these cathode materials are also highlighted.

Chapter 2 - Functionalization of 4-quinolone scaffold has attracted a great deal of attention from the synthetic community due its widespread biological activity and pharmacokinetic features. Over the last few decades, a variety of noble transition metal catalysts (mainly Pd, Cu, Rh, Ru, and others) have been fostered, enabling for the regioselective and atom-economical stitching of numerous functional groups on this potent moiety. In this perspective, the authors intend to provide a broader picture on the various transformations and concomitant annulation events for the divergent synthesis of valuable substituted 4-quinolones.

Chapter 3 - Exchange interaction at the bulk of transition metal is well described with Hund's J parameter. For bare Coulomb interactions, J can be analytically obtained. However, the value is too large. Although J is often treated as independent to the orbital-pairs, Racah parameters suggest that the value depends on the orbital pairs. In this chapter, the authors modify the Racah parameters using a Yukawa-type screened Coulomb interaction and show that in a strong screening limit, J is independent of the orbital pair.

On the other hand, the exchange parameter at the interface of transition metal is often described with s-d exchange interaction. While interfacial interaction seems to be limited only by the dirtiness of the interface, the authors show that in strong screening limit, the exchange interaction at the interface is limited by the electron-electron repulsive interaction.

Chapter 4 - At ambient temperature, a new fluorinated β-diketone ligand, 1,4-bis-(3-hexafluoroacetylacetonate)benzene (L), was synthesised and utilised to build a metal-organic framework, ML, with Cu(II) ions. Both L and ML were found to be soluble in a wide variety of solvents including tetrahydrofuran, dichloromethane, chloroform, ethyl acetate, acetone, and isopropanol. Thus, the ML was extensively characterized by infrared spectroscopy, powder X-ray diffraction, cyclic voltammetry, thermal analysis, field emission scanning electron microscopy, energy dispersive spectroscopy, atomic force microscopy, and electron paramagnetic resonance spectroscopy. The 4-coordinated Cu(II) ions of ML have been shown to be useful for electrochemical caffeine detection in preliminary electrochemical investigations.

Chapter 5 - Metal salts were efficient catalysts in different reactions such as esterification of the terpene alcohols with acetic acid and glycerol acetalization with acetone. Herein, the recent developments obtained in these two reactions using commercial metal salts as the catalysts and renewable origin alcohols (i.e., glycerol and terpene alcohol) as the substrates were discussed. Esters of terpene alcohols and glycerol acetals are valuable ingredients for fragrance, agrochemicals, and pharmaceutical industries. In all the reactions, high conversions and selectivity toward the goal products were achieved. The different aspects involved in these reactions such as temperature, catalyst load, time, and stoichiometry of reactants were assessed. The actuation of the transition metal catalysts as Lewis's acids was discussed in detail.

Chapter 1

Transition Metal Based Cathode Materials for Lithium-Ion Cells

M. Akhilash[1,2]
P. S. Salini[1]
Bibin John[1,*]
S. Sujatha[1]
and T. D. Mercy[3]

[1]Energy Systems Development Division, PCM Entity, Vikram Sarabhai Space Centre, Thiruvananthapuram, India
[2]University of Kerala, Thiruvananthapuram, Kerala, India
[3]Energy Systems Group, PCM Entity, Vikram Sarabhai Space Centre, Thiruvananthapuram, India

Abstract

Lithium-ion batteries (LIBs) are considered to be the most promising power source for electric vehicles (EVs) and portable electronic devices. Transition metals are the cornerstone of cathode materials, the key component that determine the overall performance of LIBs. This chapter reviews the advances in exploring and developing various transition metal oxides as high-performance cathode materials for LIBs. The primary focus is on the structure and especially the electrochemical properties of several classes of transition metal-based cathodes. Moreover, the strategies capable of overcoming the current limitations of these cathode materials are also highlighted.

---

* Corresponding Author's Email: bbnjohn@yahoo.com

In: Transition Metals - An Overview
Editor: Veronica J. Avila
ISBN: 979-8-88697-646-5
© 2023 Nova Science Publishers, Inc.

1. Introduction

With the severe energy crisis and environmental pollution, the development of advanced energy storage devices has received significant attention due to the intermittent nature of renewable energy sources such as solar, wind, and tidal energy.[1] Lithium-ion batteries (LIBs) are considered as one of the elements of high-tech development owing to their excellent performance such as high voltage, high energy density, no memory effect, minimal self-discharge, and satisfactory cycle life.[2,3] A schematic representation of the working principle of a lithium-ion (Li-ion) cell is shown in Figure 1.[4] The cell consists of a cathode, anode, electrolyte, and separator. Lithiated metal oxides are used as cathodes and graphite/carbon is used as the anode. An electrolyte consists of a lithium salt (e.g., $LiPF_6$) dissolved in organic solvents (e.g., Ethylene carbonate, diethyl carbonate etc.). The cathode undergoes oxidation during cell charging, releasing electrons and Li^+ ions. The Li^+ ions travel through the electrolyte to the anode, where they eventually intercalate in the layers of graphite, while the electron travels through the circuit towards the anode (Equations 1-3). When discharge occurs, the opposite reaction happens.

The energy output of Li-ion cells is limited by their positive electrode (cathode). Cathodes are essentially lithium 3d-transition metal (TM) oxides, $LiMO_2$ (M= transition metal). The efficiency of a Li-ion cell to store charge is highly influenced by the redox capacity of the TM ions. Highly efficient cathode materials with profound energy storage can be accomplished if both the TM and oxygen ions could take part in redox reactions.[5] The preferred cathode material for commercial LIBs has continued to be layered $LiMO_2$ materials. The TMs are at the center of the redox process in traditional layered oxides, which involves some degree of rehybridization between the metal d orbitals and the O 2p orbitals in conjunction with electron extraction from the TMs. [6,7] Additionally, transition metals play a significant role in lithium-ion kinetics and are highly responsible for the performance of Li-ion cells including thermal stability, cycle life, and rate capability. In this chapter, we focus on transition metal-based cathode materials with special emphasis given on 3d-series. We summarize the structural transformations as a consequence of various redox reactions taking place in the cathode material. Finally, we review the various strategies adopted to improve the electrochemical performance of cathode materials through transition metal doping and surface modification.

$$\text{Cathode: } LiCoO_2 \underset{\text{Discharge}}{\overset{\text{Charge}}{\rightleftharpoons}} Li_{1-n}CoO_2 + nLi^+ + ne^- \quad (1)$$

$$\text{Anode: } C + nLi^+ + ne^- \underset{\text{Discharge}}{\overset{\text{Charge}}{\rightleftharpoons}} Li_nC \quad (2)$$

$$\text{Overall reaction: } LiCoO_2 \underset{\text{Discharge}}{\overset{\text{Charge}}{\rightleftharpoons}} +C\ Li_{1-n}CoO_2 + Li_nC \quad (3)$$

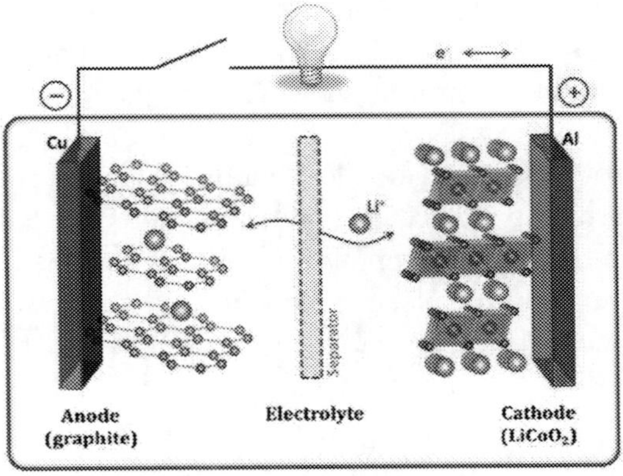

Figure 1. Schematic of the working principle of a Li-ion cell. [4] [Reprinted with permission from ref 4. Copyright 2013 American Chemical Society].

2. Cathode Materials

The performance of LIBs is greatly influenced by the cathode materials.[8] Reversible lithium insertion is possible in compounds that either have empty sites for Li-ions to occupy or have open channels (1D, 2D, or 3D) for Li-ion migration. As a result, TM complexes having open sites for Li^+ ions have been researched as lithium insertion materials. They can also exhibit varying oxidation states, which allows electron flow in the outer circuit to compensate for Li^+ ion exchange.[9,10] It has also been discovered that other materials that lack empty sites for storing lithium are also prospective candidates for LIBs. They react with lithium by a different mechanism such as alloying, conversion, etc. that is different from that of insertion.

2.1. Role of Transition Metals in Cathode Materials

Compounds selected for cathodes in LIBs are oxides of 3d-transition metals such as chromium (Cr), manganese (Mn), iron (Fe), cobalt (Co), nickel (Ni) or copper (Cu), which can change valence to maintain neutrality. The role of the TMs in the cathode is to compensate for the charge when the Li^+-ion arrives or departs. [11] For example, in the case of $LiCoO_2$ (LCO) when the Li^+ ion is taken out of the cathode structure, the cobalt in the compound changes its oxidation state (since the Li^+-ion has a positive charge), thus the oxide stays electrically neutral.[12] The electrochemical properties are largely determined by each transition metal in the cathode. First, i) the Ni in the material contributes to a high discharge capacity. However, due to undesirable reactivity with the electrolyte, rapid capacity fading is seen when the Ni level is more than 60%. ii) The Co in the material contributes to structural stability and rate capability of the cathode. Finally, iii) the Mn contributes to increasing the thermal stability and capacity retention. [13] In addition to this, 4d and 5d transition metals are also employed in LIBs as dopants. The following sections elaborately explain the electrochemical performances of various TM oxide-based cathode materials.

2.2. Types of Cathode Materials

2.2.1. Layered Cathode Materials

Layer structured α-$NaFeO_2$ type $LiMO_2$ (M = V, Mn, Co, Ni, etc.) materials have been extensively studied as cathode materials in LIBs. These materials possess a 2D interstitial site, which makes it possible to extract and insert Li^+ ions. Among the various layered materials, $LiCoO_2$, $LiNiO_2$, and their mixed oxides $Li(Ni,Co)O_2$ are extensively used for commercial applications owing to their excellent electrochemical performance and durability. The following sections briefly explain the structure, properties, electrochemical behaviour, challenges, and different strategies to enhance the electrochemical performance of various layered transition metal (TM) oxides.

2.2.1.1. Lithium Cobalt Oxide ($LiCoO_2$)

In 1980, John B. Goodenough first introduced $LiCoO_2$ (LCO) as a practical layered rock-salt structure cathode material [14,15] and efficaciously employed in commercial LIBs by Sony Corporation in 1991. LCO remains as the most exploited commercial cathode for LIBs owing to its high capacity, high structural stability during cycling, and facile synthesis in bulk quantities.[6] LCO possesses an α-$NaFeO_2$ structure, in which oxygen atoms form a cubic close-packed (ccp) arrangement, and the Li^+ and Co^{3+} ions are occupied in the octahedral (O_h) sites, such that Li^+ and Co^{3+} are arranged in the alternate (111) planes. However, the alternate arrangement of Li^+ and Co^{3+} ions causes a slight distortion in the lattice to a hexagonal symmetry (Figure 2).[16] As a result, with cell parameters of a = 2.816 (2) Å and c = 14.08 (1) Å, LCO crystallizes in the R-3m space group. [17]

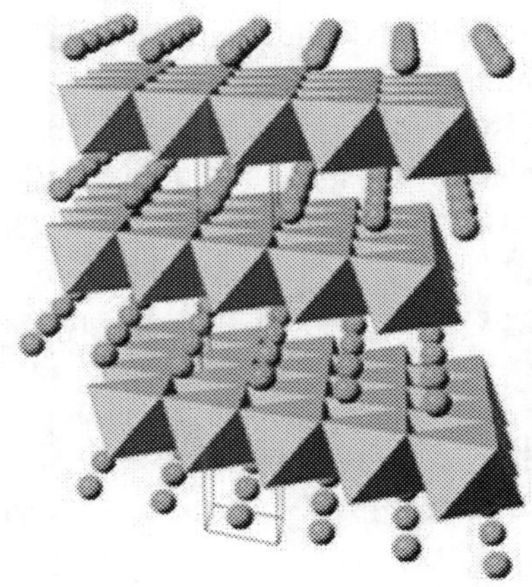

Figure 2. The crystal structure of LCO. The oxide ions form a ccp arrangement, and the Co^{3+} and Li^+ ions occupy alternating layers of octahedral sites. [16] [Reproduced from Ref. 16 with permission from the Royal Society of Chemistry].

The theoretical capacity of LCO cathode material is 274 $mAhg^{-1}$ if all the Li^+ ions are completely removed from the lattice. However, practically only about 50% of the Li^+ in the LCO structure can be reversibly used in the voltage window of 3-4.2 V, leading to a capacity of ~140 $mAhg^{-1}$.[18] The charging of the LCO cathode to a voltage of >4.2 V is restricted due to oxygen evolution. The Co^{3+}/Co^{4+} redox couple in LCO is pinned above the O-2p band and any further extraction of Li^+ ion leads to the oxidation of surface oxide ion (O^{2-}) to superoxide (O_2^-) and thus oxygen evolution takes place according to equation 4 and this will increase the safety concern.[19] Furthermore, when LCO is charged above 4.2 V, the cycling efficiency and capacity of LCO-based cells decay rapidly due to the structural instability caused by the extraction of more than 0.5 mole lithium from the lattice.

$$(O_2)^{2-} \longrightarrow O^{2-} + \tfrac{1}{2} O_2 \uparrow \qquad (4)$$

2.2.1.1.1. Alternative Cathodes for $LiCoO_2$

The high cost, less availability, and toxic nature of cobalt have pushed the research toward other transition metal-based alternative cathodes such as manganese, and nickel-based oxides with a structure similar to that of $LiCoO_2$. In this context, $LiNiO_2$ (LNO) and $LiMnO_2$ (LMO) have received much attention due to their low cost, abundance, and less toxicity.[20,21] $LiNiO_2$ was first synthesized by Dyer and his team in 1954, and later it was widely investigated as a substitute for LCO by Dahn and his co-workers in the 1990s.[22] LNO is isostructural with LCO and it allows easy intercalation of lithium ions. The main disadvantage with $LiNiO_2$ is that it is very difficult to synthesize stoichiometric $LiNiO_2$ and it always exists as $Li_{1-y}Ni_{1+y}O_2$. This is because the similar ionic radii of Ni^{2+} (0.69Å) and Li^+ (0.76Å) cause cation exchange in the material. Thus, a small amount of Ni^{2+} is always present in the lithium layer and this will cause serious difficulty in the diffusion of Li^+ during cycling. Furthermore, multiple structural changes were observed for LNO material during cycling, that is, from layered to monoclinic and then to hexagonal, and this was confirmed by Ceder et al. using first principle calculations. These multiple structural changes cause great stress on the material and ultimately lead to a severe capacity loss in the material. [23]

In the case of $LiMnO_2$ cathode material also it is very difficult to synthesize them through normal ceramic methods or other high-temperature synthesis routes.[24] Therefore, $LiMnO_2$ is usually prepared by the cation exchange reaction of $NaMnO_2$. In the $LiMnO_2$ cathode, manganese exists in the Mn^{3+} state. The theoretical capacity of $LiMnO_2$ material is 285 mAh g^{-1}, and practically it delivers a specific capacity of 190 mAh g^{-1} in the voltage range between 2.00 V and 4.25 V (vs. Li^+/Li). However, the obtained $LiMnO_2$ material undergoes a phase transition from layered to spinel structure during the cycling process and this is mainly owing to the Jahn-Teller distortion associated with the Mn^{3+} ion.[25] Thus, a partial replacement of Ni or Mn with Mg, Co, Cr, Fe, or Ti in LMO and LNO has been investigated to improve the electrochemical performance of both $LiMnO_2$ and $LiNiO_2$ cathodes. It has been reported that this partial replacement stabilizes the crystal structure during the de-intercalation of Li^+-ions.[26] These modifications eventually led to the development of ternary cathode materials, exhibiting a capacity value higher than conventionally employed $LiCoO_2$ cathode material. The layered oxides of Ni, Mn, and Co (abbreviated as NMC or NCM) are found to be more stable among several such cathode materials and are already commercialized. [5]

2.2.1.2. Lithium Nickel Manganese Cobalt Oxide (NMC) Cathode Materials

In comparison with LCO, NMC cathode material delivers robust overall performance, outstanding specific capacity, high safety, and the lowest self-heating rate of all mainstream oxide cathode materials and this makes them a better option for automotive batteries. [11] Performances associated with each transition element in NMC are higher capacity (nickel), improved safety (manganese), and better rate capability (cobalt). Generally, the change in the oxidation state of nickel (ie., the redox couples of Ni^{2+}/Ni^{3+} and Ni^{3+}/Ni^{4+}) will contribute to the majority of the reversible electrochemical capacity of the material.The oxidation state of manganese will mostly remain at +4 state during cycling and thus acts as a structural stabilizer. The presence of Co will facilitate the formation of stoichiometric compounds with little cation mixing, which will increase the material's structural stability over prolonged cycling.[27, 28]

Layered NMC 111 or NMC 333 ($LiNi_{0.33}Mn_{0.33}Co_{0.33}O_2$) is popularized as one amongst the most appealing cathode candidate for commercial LIBs.[29] Compared with other NMC cathode materials, NMC 111 is easier to synthesize and exhibits excellent cycling performance and structural stability. NMC 111 material is composed of Mn^{4+}, Co^{3+}, and Ni^{2+} ions. Studies revealed that NMC 111 material possesses less occupation of Ni^{2+} in the Li layer (that is, low cation exchange) and better reversibility. Furthermore, Co in the NMC 111 cathode enhances their properties like electrical conductivity, thereby enhancing the high-rate capability. Cycling studies of NMC 111 cathode at a working voltage window of 2.5–4.3 V deliver a specific capacity of ~150 mAh g^{-1}. Yabuuchi and Ohzuku reported that cycling of NMC 111 cathode at a higher voltage window of 2.5-4.6 V delivers capacities of ~ 200 mAh g^{-1} but at the cost of cyclability.[30] Later, Li and his team synthesized a porous nano-micro hierarchical NMC 111 microspheres via a simple nano-etching-template route, and the cathode delivered a high specific capacity even at 1 C rate (133.2 mAh g^{-1}). Nevertheless, the transition metal dissolution, poor electronic conductivity, and side reactions between NMC 111 and electrolyte led to poor cycling stability and rate performance, and this seriously plagues its large-scale applications. For cathode materials, it is critical to attain better reversible capacity and greater cycling stability owing to the growing demand for higher energy density.

Recently, the nickel-rich $LiNi_{1-x-y}Co_xMn_yO_2$ ($1-x-y \geq 0.5$) cathode materials with various chemical compositions have drawn considerable research interest because of their high specific energy (800 $Whkg^{-1}$), low cost, and environmental friendliness. Generally, Ni-rich cathode materials possess a α-$NaFeO_2$ structure (R-3m space group) analogous to $LiNiO_2$ cathodes, as they are originated by the partial cation replacement for Ni atoms in $LiNiO_2$. In these materials, the alternate layers are formed by the octahedron of LiO_6 and MO_6 with a shared edge, where the Li^+ ions occupy the 3a site, M ions occupy the site of 3b sites, and O^{2-} ions occupy the 6c site.[31] The Li and TM ions occupy O_h sites in alternate layers with the ABCABC... arrangement, which is denoted as "O3 type" (Figure 3).[32] The Ni-rich layered oxides are regarded as one of the potential materials for EV applications owing to their high specific capacities (more than 200 mAh g^{-1} at high voltages of 4.3–4.6 V) and low industrial cost. Several promising Ni-rich materials for high-energy-density LIBs have been developed including NMC 433, NMC 532, NMC 622, and NMC 811. In electric vehicles, NCM-532 cathodes are employed, which exhibit better structural stability during the Li^+ ion extraction/insertion process and also delivered a reasonable discharge capacity of ~160 mAh g^{-1}.[33] To enhance the specific capacity at acceptable upper cut-off voltages, higher Ni contents are necessary; this can be done by using Ni-rich NMCs like NMC-811 with a higher Ni content. At reasonable cut-off potentials (4.2 V vs. graphite), these Ni-rich NMCs can deliver specific capacities >180 mAh g^{-1}.[34]

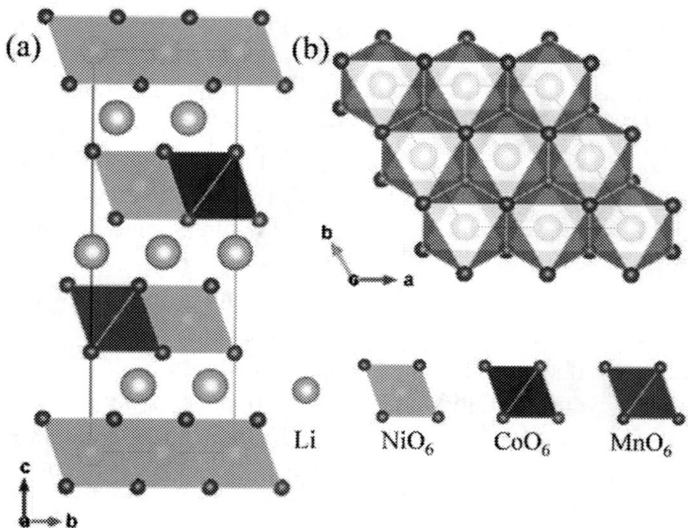

Figure 3. Well-ordered layered material with R-3m structure projected onto the (a) *bc* and (b) *ab* planes. [32][Reproduced with permission from Ref. 32]

Recent studies revealed that Ni-rich cathode materials show some serious problems for commercialization, such as poor thermal stability, high residual lithium, poor rate capability, capacity fade during long-term cycling, short storage lifetime at elevated temperature, Li/Ni cation mixing, gas evolution, serious safety issues, and so on. Therefore, researchers proposed some crucial mechanisms to clarify these thorny issues: (1) dissolution of the TM ions by hydrofluoric acid present in the electrolyte; (2) side reaction with electrolyte catalyzed by the delithiated Ni-rich cathode at voltages greater than 4.3 V with an associated oxygen release; (3) transformation of layered phase to spinel and the formation of NiO-type rock salt phase during cycling; (4) when charged and discharged repeatedly, internal strain causes the lattice volume to expand and contract, resulting in micro crack development and particle fracture. To solve these serious issues various modifications such as elemental doping and surface coating methods were adopted. Doping with elements will enhance the structural stability of the cathode and surface modification reduces the side reaction between the cathode materials and the electrolyte.

2.2.1.3. Lithium-Rich Cathode Materials

Recently, lithium-rich layered metal oxides have aroused great attention as potential candidates for next-generation LIBs due to their high specific capacities (>250 mAh g^{-1}), low cost, and low toxicity (Li-rich cathode materials are mainly composed of manganese and a small amount of cobalt and nickel).[35,36] In 1993, Thackeray and his team [37] synthesized $Li_{1.09}Mn_{0.91}O_2$ cathode by treating Li_2MnO_3 with acid and then relithiating it, and they also popularized the concept of $xLi_2MnO_3.(1\text{-}x)LiMO_2$, which is a representation of Li-rich cathode material. Later, Kalyani and team [38] found that Li_2MnO_3 material could be electrochemically cycled. Lu and his co-workers [39,40] put these ideas together and synthesized a series of $Li[Li_{(1/3-2x/3)}Mn_{(2/3-x/3)}Ni_x]O_2$ materials and studied their electrochemical performance in detail. In the subsequent years, research interest in this topic expanded and a plenty of work focusing on enhancing the electrochemical performance of these materials was carried out by incorporating other transition and non-transition metal ions into the crystal structure or by surface coating. Li-rich cathode materials are generally represented by three notations. The first two are xLi[$Li_{1/3}Mn_{2/3}$]O_2.(1− x)$LiMO_2$ and $x$$Li_2MnO_3$ (1 − x)$LiMO_2$ respectively, while the third one is $Li_{1+x}M_{1-x}O_2$, where M= Ni, Co, Mn, etc. For instance, $0.6Li[Li_{1/3}Mn_{2/3}]O_2\cdot 0.4LiMO_2$ can be represented as $0.5Li_2MnO_3\cdot 0.5LiMO_2$ (where, M = [$Mn_{1/2}Ni_{1/2}$]), and is equivalent to $Li_{1.2}Mn_{0.6}Ni_{0.2}O_2$. This is one of the widely investigated cobalt-free Li-rich cathode materials in LIBs. [41]

Generally, a Li-rich cathode material is composed of $LiMO_2$ and Li_2MnO_3 phases. $LiMO_2$ possesses a rhombohedral structure with an R-3m space group, while Li_2MnO_3 has a monoclinic structure with C2/m space group (Figure 4).[42] Both Li_2MnO_3 and $LiMO_2$ phases possess a ccp oxygen array in which all the resulting octahedral (O_h) voids are occupied by alternating layers of Li and TM or Li/TM.[43,44] XRD pattern of a Li-rich $Li_{1.5}Ni_{0.25}Mn_{0.75}O_{2.5}$ cathode material is shown in Figure 5. All the diffraction peaks in the XRD pattern of this Li-rich cathode are well-indexed to the R-3*m* space group, with some weak reflections observed in the 20–25° range. These superlattice reflections arise as a result of Li^+/Mn^{4+} ordering in the TM layer, and they, along with the major peaks, are indexed to the *C*2/*m* space group. [45,46]

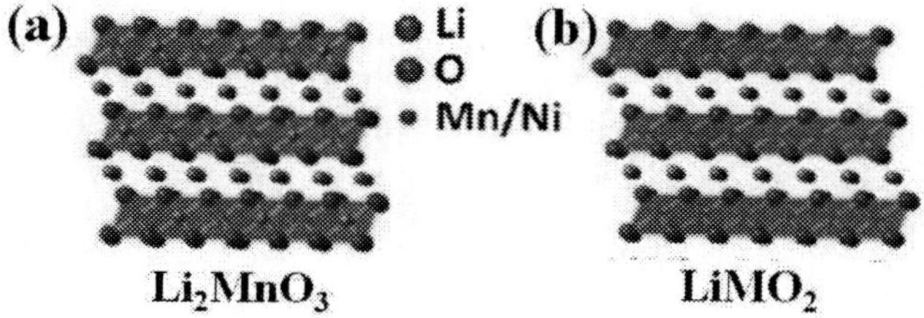

Figure 4. Crystallographic structure of (a) Li_2MnO_3 phase with *C*2/*m* symmetry, (b) $LiMO_2$ phase with R-3*m* symmetry [42][Reproduced with permission from Ref. 42.Copyright © 2016 American Chemical Society]

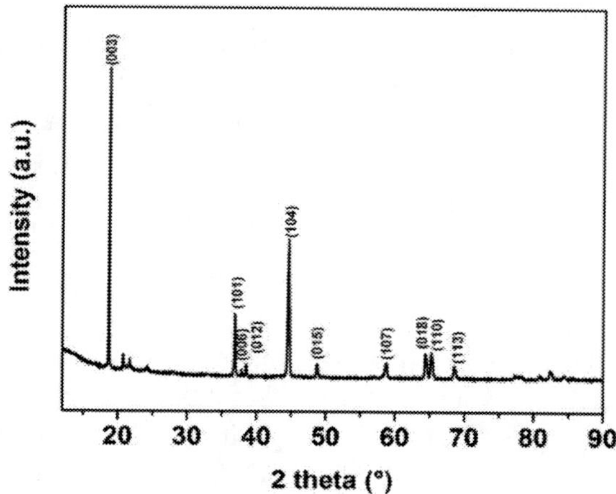

Figure 5. XRD pattern of a typical Li-rich cathode material having composition $Li_{1.5}Ni_{0.25}Mn_{0.75}O_{2.5}$ [45][Reproduced with permission from Ref. 45. Copyright ©2021 Elsevier]

Lithium-rich layered cathode materials exhibit a unique initial charge-discharge cycle mechanism, which still needs to be clarified. The first charge cycle of lithium-rich cathode material consists of two regions: (1) a region below 4.5 V (sloping region) and (2) a long plateau above 4.5 V (oxygen-loss plateau). The region below 4.5 V arises due to the extraction of lithium-ion with the simultaneous oxidation of TM ion into a +4 oxidation state in the $LiMO_2$ phase. The second region above 4.5 V is mainly due to the activation of the Li_2MnO_3 phase in the crystal structure. When the Li-rich material is charged to above 4.5 V, there is a concomitant loss of Li^+ and O^{2-} (in the form of Li_2O) from the Li_2MnO_3 component and it is converted into an electrochemically active MnO_2 phase. The TM ions that were oxidized during charging are reduced back to their original states during the initial discharge. Then another ion must be reduced due to the loss of O^{2-} ions during initial charging, which causes some of the Mn^{4+} ions to be reduced to Mn^{3+}.[47–49] This mechanism can elucidate the high specific capacities associated with Li-rich cathode materials (up to 300 mAh g^{-1}).

Among the various lithium-rich cathode materials, $Li_{1.2}Ni_{0.2}Mn_{0.6}O_2$ and $Li_{1.2}Mn_{0.54}Ni_{0.13}Co_{0.13}O_2$ are the extensively investigated materials due to their superior electrochemical performances. Powder metallurgy, followed by various cooling processes, was used by Cai et al. [50] to synthesize $Li_{1.2}Ni_{0.2}Mn_{0.6}O_2$ material with varying degrees of structural imperfections. The detailed analysis revealed that the disorder of the TM layer in Li_2MnO_3 boosts its electrochemical activity, while the Li/Ni sites of the Li layer preserve its local structural stability. The electrochemical performance studies indicate that the cathode with an optimal number of structural defects delivered an initial discharge capacity of 188.2 mAh g^{-1} and retained a capacity of 144.2 mAh g^{-1} after 500 cycles with capacity retention of 76.6% at 1C. However, the material without any structural defects delivered only a capacity of 40.8 mAh g^{-1} at the end of 500 cycles. Wang and his team [51] successfully prepared a Li-rich layered oxide having the composition $Li_{1.2}Mn_{0.54}Ni_{0.13}Co_{0.13}O_2$ via the co-precipitation method. The material with olive-like morphology delivered an initial discharge capacity of 297.0 mAh g^{-1} and also exhibits improved cyclability (95.4% capacity retention after 100 cycles at 0.5 C), as well as rate capability (142.8 mAh g^{-1} at 10 C). Our group [48] recently synthesized a novel lithium-rich compound having composition $Li_{1.17}Ni_{0.34}Mn_{0.5}O_2$ via. a simple carbonate co-precipitation method. The material possesses a porous morphology, which is evident from SEM analysis (Figures 6a and 6b) and it exhibited superior electrochemical performances. At C/10 rate the material delivered an initial discharge capacity of 250 mAh g^{-1} and at the end of 200 cycles the material exhibited capacity retention of more than 90% at 1C rate (Figure 6c).

Figure 6. (a,b) FE-SEM images and (c) cycling performance of $Li_{1.17}Ni_{0.34}Mn_{0.5}O_2$ cathode material [48][Reproduced with permission from Ref. 48. Copyright © 2022 American Chemical Society].

Several major stumbling blocks hindering these attractive Li-rich layered oxide cathodes have still to be overcome for their use in practical application, viz. (1) high irreversible capacity in the initial cycle due to the extraction of Li^+ and O^{2-} from the Li_2MnO_3 phase when the material is charged above 4.5 V [52], (2) severe voltage decay and the poor cycling performance due to unsatisfactory structural stability at the high working voltage (that is, layered to spinel phase transformation) [53], and (3) poor rate capability originated from the low ionic/electronic conductivities of Li_2MnO_3 component.[54] To mitigate these shortcomings, some strategies including surface modification and TM-ions/anions doping have been employed. The structure of cathode materials can be stabilized through doping with suitable elements, which can also alter the lattice parameters and affect the diffusion coefficient of Li^+.[55] The modification of Li-rich cathode materials by surface coating with metal oxides, phosphates, and fluorides separates the active material and electrolyte, thus enhancing thermal stability and alleviating capacity loss. [56]

2.2.2. Olivine Cathodes

$LiFePO_4$ (LFP) is the most important cathode material in this class. LFP has been widely investigated for over two decades and has been chosen as one of the strongest candidates for EV applications due to its high safety, superior thermal and chemical stability, and long cyclability.[57–59] In 1977, Yakubovich analyzed its crystal structure for the first time. [60] The orthorhombic lattice structure of the triphylite LFP, a member of the olivine family of lithium ortho-phosphates, has a Pnma space group. [61,62] The lattice parameters are a = 10.33 Å, b = 6.01 Å, c = 4.69 Å and V = 291.2 A^3. Corner-shared FeO_6 and edge-shared LiO_6 octahedra parallel to the b-axis forms the crystal structure, and PO_4 tetrahedra link them together. (Figure 7).[63] Electrochemical studies show that LFP delivered a practical capacity of 170 mAh g^{-1} (close to the theoretical capacity value) since one mole of Li^+ can be extracted per mole of LFP. The electrochemical reactivity of LFP was first recognized by Padhi and his team [61] in the year 1997. They proved that upon delithiation, Li^+ can be extracted electrochemically from LFP thus leaving hetero site iron phosphate ($FePO_4$) without changing the olivine framework. The lattice parameters are changed to a = 9.81 Å, b = 5.79 Å, c = 4.78 Å, and V = 271.5 A^3 for $FePO_4$, which corresponds to a decrease in a and b by 5% and 3.7%, an increase in c by 1.9%, and a decrease in lattice volume by 6.77%, respectively. The lithium intercalation and deintercalation process progress through a two-phase process and the ordered olivine framework is retained during cycling. [64] Therefore, LFP is capable of cycling for thousands of cycles without capacity loss. In LFP, the oxygen atoms are strongly bonded by both Fe and P atoms and thus, its crystal structure is more stable than layered oxides such as LCO at high temperatures. In the delithiated state, LFP is stable up to 400 °C, whereas LCO decomposes at 250 °C.[65,66] Thus, the excellent cyclability and safe operation of LFP are made possible by the high lattice stability.

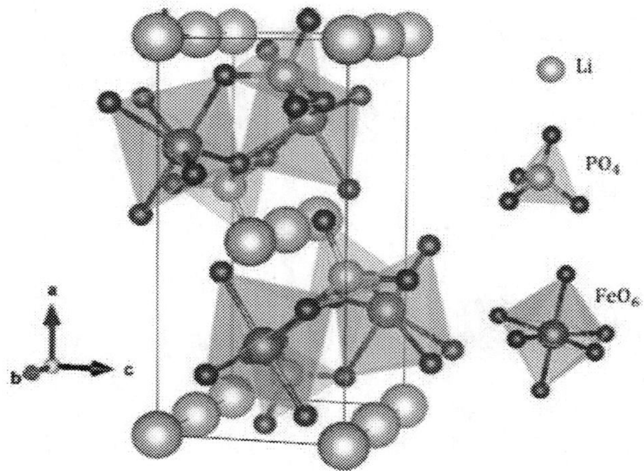

Figure 7. Crystal structure of $LiFePO_4$ [63] [Reproduced with permission from Ref. 63].

The electronic conductivity of LFP is in the range of 10^{-9}-10^{-10} Scm^{-1}, which is very low and is owing to the strong covalent oxygen bonds. [67] Li^+ diffusion through LFP is widely believed to restrict to tunnels along the 'b'-axis and hence are 1D ionic conductors. Slow Li^+ ion diffusion and poor electronic conductivity of LFP require a much slower charge-discharge rate process to obtain the full capacity. At a high current rate, extraction/insertion is restricted to 0.6 lithium. Nevertheless, the conductivity and rate capability of LFP can be improved by (i) coating with conducting carbon (ii) creating small particles in the nano regime, and (iii) elemental doping.

2.2.3. Spinel Cathodes

Spinel $LiMn_2O_4$ material is one of the most attractive and most studied cathode materials for rechargeable LIBs due to the availability of Mn, easy synthesis method, low production cost, non-toxicity, and high-rate capability. [68,69] This material also exhibits high thermal stability. It has a theoretical capacity of 148 mAh g^{-1}. $LiMn_2O_4$ material crystallizes in spinel structure with cubic symmetry and it belongs to the Fd3m space group [70,71] and is shown in Figure 8. [16] In $LiMn_2O_4$ structure, the Li^+ ions are occupied in 8a tetrahedral sites and Mn^{3+}/Mn^{4+} ions are located in 16d octahedral sites, whereas O^{2-} are placed in 32e sites, resulting in a ccp array. [72,73] A diamond-shaped framework is created by the $[MnO_6]^{2-}$ octahedra, which offers a 3D channel for Li^+ ion movement during material cycling. [71,74] In comparison to

$LiMn_2O_4$ cathodes, layered LCO and LNO provide a 2D diffusion path, whereas LFP furnishes a 1D Li^+ ion diffusion path. [75] Recently, Yang et al. [76] synthesized $LiMn_2O_4$ spinel with 1D hollow structure via. single-spinneret electrospinning. The major advantages of this 1D $LiMn_2O_4$ structure are that it will provide a large surface area, more diffusion paths for lithium ions, and a low risk of unit cell change during the Li^+-ion insertion/de-insertion processes. [76]

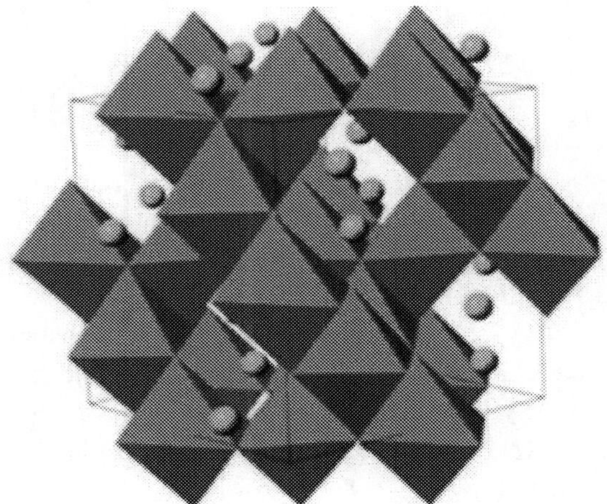

Figure 8. Structure of spinel $LiMn_2O_4$ showing the existence of 3D channels for lithium diffusion [16][Reproduced from Ref. 16 with permission from the Royal Society of Chemistry].

During initial charging, lithium ion moves from one 8a tetrahedral site to another 8a tetrahedral site through the 16c octahedral site. However, the movement of lithium ions through 16c octahedral sites is unfavourable and requires higher energy and thus, the extraction of Li^+ ions take place at the 4 V region. Li^+-ions are inserted back into the spinel structure during discharge in a similar reverse process, which is shown in equation 5.[77] 16c site of $LiMn_2O_4$ is the unoccupied, and additional insertion of Li^+ into crystal structure is possible during discharge to form $Li_{1+x}Mn_2O_4$ (equation 6). This happens in the low voltage region of 3V and thus, $LiMn_2O_4$ is capable of inserting ~2 Li^+ ions into the structure. Nevertheless, intercalation of more Li^+ ions into the spinel structure increases the number of Jahn-Teller ions (Mn^{3+}: $t_{2g}^3e_g^1$) owing to the decrease in average oxidation state from 3.5 to 3.0 and causes the cooperative distortion of MnO_6 octahedra. Hence, cycling up to 3

V causes the phases transition, that is, from spinel phase totetragonal phase with large deviation in unit cell volume and this will lead to poor cycling of $LiMn_2O_4$. [78,79] Al_2O_3 coating, substitution of manganese with various metal ions such as $M^+(Li^+)$, $M^{2+}(Zn^{2+}, Mg^{2+})$, M^{3+} $(Cr^{3+}, Co^{3+}, Al^{3,+})$ and $M^{4+}(Ti^{4+})$, etc, and stabilizing Li excess phase were used to enhance the structural stability of the $LiMn_2O_4$ cathode.

$$LiMn_2O_4 \rightleftharpoons xLi^+ + Li_{1-x}Mn_2O_4 + xe^- \quad (5)$$

$$LiMn_2O_4 + xLi^+ + xe^- \rightleftharpoons Li_{1+x}Mn_2O_4 \quad (6)$$

2.2.4. Inverse Spinel

The new class of materials such as $LiNiVO_4$, $LiCoVO_4$, $LiCo_{1-x}Ni_xVO_4$ (where x = 0.5), etc., with inverse spinel structure, are potential cathode candidates for LIBs owing to their high working voltages of 4.2-5.0 V vs. Li^+/Li with a specific capacity of 148 mAh g^{-1}. [80] The voltage difference between $LiCoVO_4$ (4.2 V) and $LiNiVO_4$ (4.8 V) is substantial which implies that the presence of Ni atoms in both inverse spinel and spinel structures play a crucial role in the voltage behaviour in these kinds of materials. Based on X-ray diffraction (XRD) and Rietveld refinement analysis, Fey and his team [81] proved that both orthovanadates have an inverse spinel structure, where the cation arrangement was $(V)^{IV}$ $(LiM)^{VI}O_4$. These materials have lower capacity compared to other conventionally employed materials and thus, efforts were made to improve the performance of these cathode materials via. partial substitution of TM ions by other suitable metal cations. [82]

2.3. Strategies to Enhance Electrochemical Performance of Cathode Materials

2.3.1. Surface Modification

In cathode materials, the observed capacity loss arises from extrinsic factors, such as decomposition of surface carbonates or TM ion dissolution, besides due to some intrinsic loss mechanisms. Upon de-lithiation, changes are occurring in the chemical bonding between the TM and oxygen, which destabilize the cathode candidate through various degradation mechanisms, including phase transitions and O_2 evolution at the surface, leading to capacity fading.[83,84] To stabilize the crystal structure, researchers adopted surface modifications in cathode materials by coating various metal oxides, phosphates, and fluorides. Coatings form a physical barrier over the secondary particle and hence allegedly stabilize the cathode structure by hindering O_2 release, thus protecting it from impurities such as hydrofluoric acid. [85,85] Apart from this, coatings often provide additional advantageous effects. Due to their innately low ionic conductivity and decomposition with gassing, residual Li in the cathode materials has a detrimental effect on the electrochemical performance of these materials. In addition to this, some coatings convert this residual Li into lithium compounds, such as garnets and phosphates thereby improving the conductivity which in turn enhances the rate capability. [86,87] Studies reveal that polymeric coatings can completely cover surfaces more readily, and by providing secondary particles with elastic support, this will help to prevent cracking.[88] The surface coating also hinders TM dissolution and improves the thermal stability of the cathode material during cycling. [89] In the following section, we list some of the most commonly employed coatings.

2.3.1.1. Oxides

The strong covalent bonds in oxides forms a stable defensive layer to safeguard the cathode surface from undesirable reactions with the electrolyte, though low ionic and electronic conductivity of these oxides causes poor kinetics and high impedance. This problem can be resolved to some extent by reaction with lithium compounds as well as calcination, especially at the surface of Li-rich and Ni-rich cathode materials. The obtained mixed oxides have improved ionic conductivity and also maintain the stability as the starting oxides. Moreover, oxide based coatings can operate both as an HF barrier and scavenger. [90] Typical examples are TiO_2 [91], SiO_2 [92], Al_2O_3 [93], and ZrO_2 [94]. Oxide coatings based on boron [95],tungsten [96], cobalt [97], and tin [98] have also been reported.

2.3.1.2. Fluorides

Fluoride coatings act as a thermodynamically stable barrier between the electrolyte and the cathode material and are generally achieved by mixing NH_4F and respective metal nitrates [99], or by using the atomic layer deposition (ALD) method [100]. Commonly used fluoride coating includes LiF [101], FeF_3 [102], AlF_3 [99], and $LiAlF_4$ [100]. Coating with LiF results in shifting of equilibrium of the electrolyte decomposition reaction ($LiPF_6 \rightleftharpoons LiF + PF_5$) to the left, thus hindering hydrofluoric acid formation in the cell. Besides reduced capacity fading, the studies also reveal better rate performance for the cathodes coated with LiF. Furthermore, AlF_3 has a lower Gibbs free energy of formation than Al_2O_3, it is thought that fluorides are more stable in ambient air than oxides.

2.3.1.3. Phosphates

Phosphate-based surface modification is usually employed to obtain residual Li based compounds that exist on the surface of Li-rich and Ni-rich cathode materials. After being treated with phosphates, the electrochemically inert lithium carbonates and hydroxides, which are detrimental for long-term cycling, are transformed into an active, ionic-conducting layer which can protect the surface further. Typical examples of phosphates include Li_3PO_4 [103], $AlPO_4$ [104], and $FePO_4$ [105] and these phosphate coatings improve the processability, storage life, and electrochemical performance of cells.

2.3.2 Elemental Doping

Doping is a widely employed technique to stabilize cathode materials and is considered as an effective strategy to enhance the performance of these materials. Modification by doping, also known as substitution modification, replaces a part of the corresponding position of the original material structure with a small number of other ions or groups. The crystal and electronic structure are typically stabilized by doping cathode materials with various metal ions or anionic species because this elevates the thermodynamic or kinetic barrier for oxygen loss, which further prevents the phase transition. [106] Dopant ions with strong metal-oxygen (M-O) bonds can offer several advantages. Firstly, by preventing Jahn-Teller distortion, metallic doping creates pillars and ensures the stability of cathode materials, particularly for layered structures, during the charge/discharge process.[107] Secondly, by preventing O_2 evolution and elevating formation energies for cation disordering, Li/Ni mixing can be controlled.[108] Additionally, metallic doping reduces the thickness of SEI film by preventing side reactions at the electrode/electrolyte interface due to the high bonding energy.

Based on various states of doping ions, doping can be classified as anion doping, cation doping, and co-doping modification. Among these, cation doping is more widely employed. Cation doping involves replacing a part of lithium ions or TM ions with other metal cations, typically in TM oxides. [109] There have been several studies on the effects of introducing dopant elements (such as Al, Mg, Ti, Cr, Ga, Fe, Zn, etc.) into TM-based cathode materials. [110–112] Elements whose radius is similar to that of oxygen, such as F, Cl, S, etc., are used to replace a small amount of oxygen in TM-based oxide cathode materials. [113,114] Anion doping has also been extensively employed in polyanionic cathode materials. The polyanionic cathode has a limited specific capacity and energy density due to their high molecular weight. However, doping with anions with higher-polarization such as F^- or $(SO_4)^{2-}$ can enhance the working voltage and doping with anions with high charge-to-mass ratio (e.g., F^- or O^{2-}) is advantageous to achieve higher capacity. Co-doping modification is another strategy to improve the electrochemical performances of cathode materials. Multielement co-doping, in contrast to single-element doping, produces a synergistic effect by combining the positive effects of single-element doping. [115]

Conclusion

The cathode material is one of the vital components of LIBs, and it influences the performance and price of the final battery pack. Generally, cathode materials employed in LIBs are oxides of transition metals, which can undergo oxidation to higher valences when lithium is removed from their crystal structure. Several transition metal-based cathode materials including layered $LiCoO_2$, $LiNi_{0.33}Mn_{0.33}Co_{0.33}O_2$ (NMC 111), $LiNi_{0.6}Co_{0.2}Mn_{0.2}O_2$ (NMC 622), and $LiNi_{0.8}Mn_{0.1}Co_{0.1}O_2$ (NMC 811), olivine $LiFePO_4$, and spinel Li_2MnO_4 provide high voltages and good specific capacities. In addition to these cathodes, recently Lithium rich layered cathode materials are also gaining much attention in the battery community due to their exceptionally high discharge capacity and high working voltage. However, full utilization of these materials for several recharging cycles and at high discharge currents remains to be a challenge. For achieving long operational life, stabilization of the desired crystal structure of cathode materials is essential during delithiation, and also preventing reaction with the electrolyte is important. These thorny issues can be resolved by adopting various effective strategies including cation/anion/co-doping which stabilizes material structure and also could stimulate oxygen redox reactions, and surface engineering (protective surface coating) offers improved surface stability, and the development of high-performance mixed phase materials.

Acknowledgments

Our team greatly acknowledges the Director, of Vikram Sarabhai Space Centre, Thiruvananthapuram for granting permission to publish this paper. Akhilash M., extremely expresses his sincere gratitude to the University of Kerala for granting a Junior Research Fellowship (the University of Kerala Registration No. AcEVI(2)/718/CHE/18145/2018).

References

[1] Guan, P., Zhou, L., Yu, Z., Sun, Y., Liu, Y., Wu, F., Jiang, Y., Chu, D. Recent Progress of Surface Coating on Cathode Materials for High-Performance Lithium-Ion Batteries. *J. Energy Chem.*2020, *43*, 220–235. https://doi.org/10.1016/j.jechem.2019.08.022.

[2] Ding, Y., Cano, Z. P., Yu, A., Lu, J., Chen, Z. Automotive Li-Ion Batteries: Current Status and Future Perspectives. *Electrochem. Energy Rev.*2019, *2* (1), 1–28. https://doi.org/10.1007/s41918-018-0022-z.

[3] John, B.,Sandhya, C. P., Gouri, C. *The Role of Nanosied Materials in Lithium Ion Batteries*; CRC Press, 2018.

[4] Goodenough, J. B., Park, K.-S. The Li-Ion Rechargeable Battery: A Perspective. *J. Am. Chem. Soc.*2013, *135* (4), 1167–1176. https://doi.org/10.1021/ja3091438.

[5] Liu, S., Xiong, L., He, C. Long Cycle Life Lithium Ion Battery with Lithium Nickel Cobalt Manganese Oxide (NCM) Cathode. *J. Power Sources*2014, *261*, 285–291. https://doi.org/10.1016/j.jpowsour.2014.03.083.

[6] Hausbrand, R., Cherkashinin, G., Ehrenberg, H., Gröting, M., Albe, K., Hess, C., Jaegermann, W. Fundamental Degradation Mechanisms of Layered Oxide Li-Ion Battery Cathode Materials: Methodology, Insights and Novel Approaches. *Mater. Sci. Eng. B*2015, *192*, 3–25. https://doi.org/10.1016/j.mseb.2014.11.014.

[7] Manthiram, A., Song, B., Li, W. A Perspective on Nickel-Rich Layered Oxide Cathodes for Lithium-Ion Batteries. *Energy Storage Mater.*2017, *6*, 125–139. https://doi.org/10.1016/j.ensm.2016.10.007.

[8] Whittingham, M. S. Lithium Batteries and Cathode Materials. *Chem. Rev.*2004, *104* (10), 4271–4302. https://doi.org/10.1021/cr020731c.

[9] Winter, M., Besenhard, J. O., Spahr, M. E., Novák, P. Insertion Electrode Materials for Rechargeable Lithium Batteries. *Adv. Mater.*1998, *10* (10), 725–763. https://doi.org/10.1002/(SICI)1521-4095(199807)10:10<725::AID-ADMA725>3.0.CO;2-Z.

[10] Manthiram, A., Goodenough, J. B. Layered Lithium Cobalt Oxide Cathodes. *Nat. Energy*2021, *6* (3), 323–323. https://doi.org/10.1038/s41560-020-00764-8.

[11] Schipper, F., Erickson, E. M., Erk, C., Shin, J.-Y., Chesneau, F. F., Aurbach, D. Review—Recent Advances and Remaining Challenges for Lithium Ion Battery

Cathodes. *J. Electrochem. Soc.*2016, *164* (1), A6220. https://doi.org/10.1149/2.0351701jes.

[12] Wang, L., Chen, B., Ma, J., Cui, G., Chen, L. Reviving Lithium Cobalt Oxide-Based Lithium Secondary Batteries-toward a Higher Energy Density. *Chem. Soc. Rev.*2018, *47* (17), 6505–6602. https://doi.org/10.1039/C8CS00322J.

[13] Aryal, S., Durham, J. L., Lipson, A. L., Pupek, K. Z., Kahvecioglu, O. Roles of Mn and Co in Ni-Rich Layered Oxide Cathodes Synthesized Utilizing a Taylor Vortex Reactor. *Electrochimica Acta*2021, *391*, 138929. https://doi.org/10.1016/j.electacta.2021.138929.

[14] Mizushima, K., Jones, P. C., Wiseman, P. J., Goodenough, J. B. Li_xCoO_2 (0. *Mater. Res. Bull.*1980, *15* (6), 783–789. https://doi.org/10.1016/0025-5408(80)90012-4.

[15] Goodenough, J. B., Mizushima, K., Takeda, T. Solid-Solution Oxides for Storage-Battery Electrodes. *Jpn. J. Appl. Phys.*1980, *19* (S3), 305. https://doi.org/10.7567/JJAPS.19S3.305.

[16] Islam, M. S., Fisher, C. A. J. Lithium and Sodium Battery Cathode Materials: Computational Insights into Voltage, Diffusion and Nanostructural Properties. *Chem. Soc. Rev.*2013, *43* (1), 185–204. https://doi.org/10.1039/C3CS60199D.

[17] Gummow, R., Liles, D., Thackeray, M., David, W. A Reinvestigation of the Structures of Lithium-Cobalt-Oxides with Neutron-Diffraction Data. *Mater. Res. Bull.*1993, *28* (11), 1177–1184. https://doi.org/10.1016/0025-5408(93)90098-X.

[18] Lee, J.-I., Lee, E.-H., Park, J.-H., Park, S., Lee, S.-Y. Ultrahigh-Energy-Density Lithium-Ion Batteries Based on a High-Capacity Anode and a High-Voltage Cathode with an Electroconductive Nanoparticle Shell. *Adv. Energy Mater.*2014, *4* (8), 1301542. https://doi.org/10.1002/aenm.201301542.

[19] Manthiram, A. Materials Challenges and Opportunities of Lithium Ion Batteries. *J. Phys. Chem. Lett.*2011, *2* (3), 176–184. https://doi.org/10.1021/jz1015422.

[20] Thomas, M. G. S. R., David, W. I. F., Goodenough, J. B., Groves, P. Synthesis and Structural Characterization of the Normal Spinel $Li[Ni_2]O_4$. *Mater. Res. Bull.*1985, *20* (10), 1137–1146. https://doi.org/10.1016/0025-5408(85)90087-X.

[21] Armstrong, A. R., Paterson, A. J., Robertson, A. D., Bruce, P. G. Nonstoichiometric Layered $Li_xMn_yO_2$ with a High Capacity for Lithium Intercalation/Deintercalation. *Chem. Mater.*2002, *14* (2), 710–719. https://doi.org/10.1021/cm010382n.

[22] Dahn, J. R., Sacken, U. von; Juzkow, M. W., Al-Janaby, H. Rechargeable $LiNiO_2$/Carbon Cells. *J. Electrochem. Soc.*1991, *138* (8), 2207. https://doi.org/10.1149/1.2085950.

[23] Arroyo y de Dompablo, M. E., Ceder, G. First-Principles Calculations on Li_xNiO_2: Phase Stability and Monoclinic Distortion. *J. Power Sources*2003, *119–121*, 654–657. https://doi.org/10.1016/S0378-7753(03)00199-X.

[24] Armstrong, A. R., Bruce, P. G. Synthesis of Layered $LiMnO_2$ as an Electrode for Rechargeable Lithium Batteries. *Nature* 1996, *381* (6582), 499–500. https://doi.org/10.1038/381499a0.

[25] Armstrong, A. R., Dupre, N., Paterson, A. J., Grey, C. P., Bruce, P. G. Combined Neutron Diffraction, NMR, and Electrochemical Investigation of the Layered-to-

Spinel Transformation in $LiMnO_2$. *Chem. Mater.*2004, *16* (16), 3106–3118. https://doi.org/10.1021/cm034964b.

[26] Wang, X., Wang, X., Lu, Y. Realizing High Voltage Lithium Cobalt Oxide in Lithium-Ion Batteries. *Ind. Eng. Chem. Res.*2019, *58* (24), 10119–10139. https://doi.org/10.1021/acs.iecr.9b01236.

[27] Gilbert, J. A., Bareño, J., Spila, T., Trask, S. E., Miller, D. J., Polzin, B. J., Jansen, A. N., Abraham, D. P. Cycling Behavior of NCM523/Graphite Lithium-Ion Cells in the 3–4.4 V Range: Diagnostic Studies of Full Cells and Harvested Electrodes. *J. Electrochem. Soc.*2016, *164* (1), A6054. https://doi.org/10.1149/2.0081701jes.

[28] Cho, W., Kim, S.-M., Song, J. H., Yim, T., Woo, S.-G., Lee, K.-W., Kim, J.-S., Kim, Y.-J. Improved Electrochemical and Thermal Properties of Nickel Rich $LiNi_{0.6}Co_{0.2}Mn_{0.2}O_2$ Cathode Materials by SiO_2 Coating. *J. Power Sources*2015, *282*, 45–50. https://doi.org/10.1016/j.jpowsour.2014.12.128.

[29] Ohzuku, T., Makimura, Y. Layered Lithium Insertion Material of $LiCo_{1/3}Ni_{1/3}Mn_{1/3}O_2$ for Lithium-Ion Batteries. *Chem. Lett.*2001, *30* (7), 642–643. https://doi.org/10.1246/cl.2001.642.

[30] Yabuuchi, N., Ohzuku, T. Extended Abstract No. 122, IMLB 11, June 23-28, Monterey, CA, 2002. *Google Sch. There No Corresp. Rec. This Ref.*

[31] Liu, W., Oh, P., Liu, X., Lee, M.-J., Cho, W., Chae, S., Kim, Y., Cho, J. Nickel-Rich Layered Lithium Transition-Metal Oxide for High-Energy Lithium-Ion Batteries. *Angew. Chem. Int. Ed.*2015, *54* (15), 4440–4457. https://doi.org/10.1002/anie.201409262.

[32] Lim, J.-M., Hwang, T., Kim, D., Park, M.-S., Cho, K., Cho, M. Intrinsic Origins of Crack Generation in Ni-Rich $LiNi_{0.8}Co_{0.1}Mn_{0.1}O_2$ Layered Oxide Cathode Material. *Sci. Rep.*2017, *7* (1), 39669. https://doi.org/10.1038/srep39669.

[33] Blomgren, G. E. The Development and Future of Lithium Ion Batteries. *J. Electrochem. Soc.*2016, *164* (1), A5019. https://doi.org/10.1149/2.0251701jes.

[34] Kim, M.-H., Shin, H.-S., Shin, D., Sun, Y.-K. Synthesis and Electrochemical Properties of $Li[Ni_{0.8}Co_{0.1}Mn_{0.1}]O_2$ and $Li[Ni_{0.8}Co_{0.2}]O_2$ via Co-Precipitation. *J. Power Sources*2006, *159* (2), 1328–1333. https://doi.org/10.1016/j.jpowsour.2005.11.083.

[35] Johnson, C. S., Li, N., Lefief, C., Thackeray, M. M. Anomalous Capacity and Cycling Stability of $xLi_2MnO_3 \cdot (1-x)LiMO_2$ Electrodes (M=Mn, Ni, Co) in Lithium Batteries at 50°C. *Electrochem. Commun.*2007, *9* (4), 787–795. https://doi.org/10.1016/j.elecom.2006.11.006.

[36] Armstrong, A. R., Holzapfel, M., Novák, P., Johnson, C. S., Kang, S.-H., Thackeray, M. M., Bruce, P. G. Demonstrating Oxygen Loss and Associated Structural Reorganization in the Lithium Battery Cathode $Li[Ni_{0.2}Li_{0.2}Mn_{0.6}]O_2$. *J. Am. Chem. Soc.*2006, *128* (26), 8694–8698. https://doi.org/10.1021/ja062027+.

[37] Rossouw, M. H., Liles, D. C., Thackeray, M. M. Synthesis and Structural Characterization of a Novel Layered Lithium Manganese Oxide, $Li_{0.36}Mn_{0.91}O_2$, and Its Lithiated Derivative, $Li_{1.09}Mn_{0.91}O_2$. *J. Solid State Chem.*1993, *104* (2), 464–466. https://doi.org/10.1006/jssc.1993.1182.

[38] Kalyani, P., Chitra, S., Mohan, T., Gopukumar, S. Lithium Metal Rechargeable Cells Using Li_2MnO_3 as the Positive Electrode. *J. Power Sources*1999, *80* (1), 103–106. https://doi.org/10.1016/S0378-7753(99)00066-X.

[39] Lu, Z., MacNeil, D. D., Dahn, J. R. Layered Cathode Materials $Li[Ni_xLi_{(1/3-2x/3)}Mn_{(2/3-x/3)}]O_2$ for Lithium-Ion Batteries. *Electrochem. Solid-State Lett.*2001, *4* (11), A191. https://doi.org/10.1149/1.1407994.

[40] Lu, Z., Dahn, J. R. Understanding the Anomalous Capacity of Li/ $Li[Ni_xLi_{(1/3-2x/3)}Mn_{(2/3-x/3)}]O_2$ Cells Using In Situ X-Ray Diffraction and Electrochemical Studies. *J. Electrochem. Soc.*2002, *149* (7), A815. https://doi.org/10.1149/1.1480014.

[41] Yu, H., Ishikawa, R., So, Y.-G., Shibata, N., Kudo, T., Zhou, H., Ikuhara, Y. Direct Atomic-Resolution Observation of Two Phases in the $Li_{1.2}Mn_{0.567}Ni_{0.166}Co_{0.067}O_2$ Cathode Material for Lithium-Ion Batteries. *Angew. Chem. Int. Ed.*2013, *52* (23), 5969–5973. https://doi.org/10.1002/anie.201301236.

[42] Li, Y., Wu, C., Bai, Y., Liu, L., Wang, H., Wu, F., Zhang, N., Zou, Y. Hierarchical Mesoporous Lithium-Rich $Li[Li_{0.2}Ni_{0.2}Mn_{0.6}]O_2$ Cathode Material Synthesized via Ice Templating for Lithium-Ion Battery. *ACS Appl. Mater. Interfaces*2016, *8* (29), 18832–18840. https://doi.org/10.1021/acsami.6b04687.

[43] Akhilash, M., Salini, P. S., John, B., Mercy, T. D. A Journey through Layered Cathode Materials for Lithium Ion Cells – From Lithium Cobalt Oxide to Lithium-Rich Transition Metal Oxides. *J. Alloys Compd.*2021, *869*, 159239. https://doi.org/10.1016/j.jallcom.2021.159239.

[44] Akhilash, M., Salini, P. S., John, B., Mercy, T. D. The Renaissance of High-Capacity Cathode Materials for Lithium Ion Cells. In *Energy Harvesting and Storage: Fundamentals and Materials*; Jayaraj, M. K., Antony, A., Subha, P. P., Eds., Energy Systems in Electrical Engineering; Springer Nature: Singapore, 2022; pp 181–208. https://doi.org/10.1007/978-981-19-4526-7_6.

[45] Akhilash, M., Salini, P. S., Jalaja, K., John, B., Mercy, T. D. Synthesis of $Li_{1.5}Ni_{0.25}Mn_{0.75}O_{2.5}$ Cathode Material via Carbonate Co-Precipitation Method and Its Electrochemical Properties. *Inorg. Chem. Commun.*2021, *126*, 108434. https://doi.org/10.1016/j.inoche.2020.108434.

[46] Pillai, A. M., Salini, P. S., John, B., Pillai, S., SarojiniAmma, S., Mercy, T. Synthesis and Characterization of $Li_{1.25}Ni_{0.25}Mn_{0.5}O_2$: A High-Capacity Cathode Material with Improved Thermal Stability and Rate Capability for Lithium-Ion Cells. *J. Alloys Compd.*2023, *938*, 168363. https://doi.org/10.1016/j.jallcom.2022.168363.

[47] Pillai, A. M., Salini, P. S., John, B., Nair, V. S., Jalaja, K., SarojiniAmma, S., Devassy, M. T. Cobalt-Free Li-Rich High-Capacity Cathode Material for Lithium-Ion Cells Synthesized through Sol–Gel Method and Its Electrochemical Performance. *Ionics*2022, *28* (11), 5005–5014. https://doi.org/10.1007/s11581-022-04725-x.

[48] Mohanan Pillai, A., Salini, P. S., John, B., T, J., SarojiniAmma, S., Thelakkattu Devassy, M. Synthesis and Electrochemical Characterization of a Li-Rich $Li_{1.17}Ni_{0.34}Mn_{0.5}O_2$ Cathode Material for Lithium-Ion Cells. *Energy Fuels*2022, *36* (18), 11186–11193. https://doi.org/10.1021/acs.energyfuels.2c01231.

[49] Pillai, A. M., Salini, P. S., John, B., Devassy, M. T. Aqueous Binders for Cathodes: A Lodestar for Greener Lithium Ion Cells. *Energy Fuels*2022, *36* (10), 5063–5087. https://doi.org/10.1021/acs.energyfuels.2c00346.

[50] Cai, Z., Wang, S., Zhu, H., Tang, X., Ma, Y., Yu, D. Y. W., Zhang, S., Song, G., Yang, W., Xu, Y., Wen, C. Improvement of Stability and Capacity of Co-Free, Li-Rich Layered Oxide $Li_{1.2}Ni_{0.2}Mn_{0.6}O_2$ Cathode Material through Defect Control. *J. Colloid Interface Sci.*2023, *630*, 281–289. https://doi.org/10.1016/j.jcis.2022.10.105.

[51] Wang, G., Yi, L., Yu, R., Wang, X., Wang, Y., Liu, Z., Wu, B., Liu, M., Zhang, X., Yang, X., Xiong, X., Liu, M. $Li_{1.2}Ni_{0.13}Co_{0.13}Mn_{0.54}O_2$ with Controllable Morphology and Size for High Performance Lithium-Ion Batteries. *ACS Appl. Mater. Interfaces*2017, *9* (30), 25358–25368. https://doi.org/10.1021/acsami.7b07095.

[52] Liu, Y., Zhang, Z., Fu, Y., Wang, Q., Pan, J., Su, M., Battaglia, V. S. Investigation the Electrochemical Performance of $Li_{1.2}Ni_{0.2}Mn_{0.6}O_2$ Cathode Material with $ZnAl_2O_4$ Coating for Lithium Ion Batteries. *J. Alloys Compd.*2016, *685*, 523–532. https://doi.org/10.1016/j.jallcom.2016.05.329.

[53] Wei, Z., Xia, Y., Qiu, B., Zhang, Q., Han, S., Liu, Z. Correlation between Transition Metal Ion Migration and the Voltage Ranges of Electrochemical Process for Lithium-Rich Manganese-Based Material. *J. Power Sources*2015, *281*, 7–10. https://doi.org/10.1016/j.jpowsour.2015.01.149.

[54] He, Z., Wang, Z., Huang, Z., Chen, H., Li, X., Guo, H. A Novel Architecture Designed for Lithium Rich Layered $Li[Li_{0.2}Mn_{0.54}Ni_{0.13}Co_{0.13}]O_2$ Oxides for Lithium-Ion Batteries. *J. Mater. Chem. A*2015, *3* (32), 16817–16823. https://doi.org/10.1039/C5TA04424C.

[55] Chen, H., Hu, Q., Huang, Z., He, Z., Wang, Z., Guo, H., Li, X. Synthesis and Electrochemical Study of Zr-Doped $Li[Li_{0.2}Mn_{0.54}Ni_{0.13}Co_{0.13}]O_2$ as Cathode Material for Li-Ion Battery. *Ceram. Int.*2016, *42* (1, Part A), 263–269. https://doi.org/10.1016/j.ceramint.2015.08.104.

[56] Wu, Y., Manthiram, A. Effect of Surface Modifications on the Layered Solid Solution Cathodes $(1-z)Li[Li_{1/3}Mn_{2/3}]O_2-(z)\ Li[Mn_{0.5-y}Ni_{0.5-y}Co_{2y}]O_2$. *Solid State Ion.*2009, *180* (1), 50–56. https://doi.org/10.1016/j.ssi.2008.11.002.

[57] Ritchie, A., Howard, W. Recent Developments and Likely Advances in Lithium-Ion Batteries. *J. Power Sources*2006, *162* (2), 809–812. https://doi.org/10.1016/j.jpowsour.2005.07.014.

[58] Tarascon, J.-M., Armand, M. Issues and Challenges Facing Rechargeable Lithium Batteries. *Nature*2001, *414* (6861), 359–367. https://doi.org/10.1038/35104644.

[59] Park, O. K., Cho, Y., Lee, S., Yoo, H.-C., Song, H.-K., Cho, J. Who Will Drive Electric Vehicles, Olivine or Spinel? *Energy Environ. Sci.*2011, *4* (5), 1621–1633. https://doi.org/10.1039/C0EE00559B.

[60] Yakubovich, O. V., Simonov, M. A., Belov, N. V. The Crystal Structure of a Synthetic Triphylite $LiFe[PO_4]$. *Sov. Phys. Dokl.*1977, *22*, 347.

[61] Padhi, A. K., Nanjundaswamy, K. S., Goodenough, J. B. Phospho-olivines as Positive-Electrode Materials for Rechargeable Lithium Batteries. *J. Electrochem. Soc.*1997, *144* (4), 1188. https://doi.org/10.1149/1.1837571.

[62] Andersson, A. S., Thomas, J. O. The Source of First-Cycle Capacity Loss in $LiFePO_4$. *J. Power Sources*2001, *97–98*, 498–502. https://doi.org/10.1016/S0378-7753(01)00633-4.

[63] Astuti, F., Maghfirohtuzzoimah, V. L., Intifadhah, S. H., Az-Zahra, P., Arifin, R., Klysubun, W., Zainuri, M., Darminto. Local Structure and Electronic Structure of $LiFePO_4$ as a Cathode for Lithium-Ion Batteries. *J. Phys. Conf. Ser.*2021, *1951* (1), 012007. https://doi.org/10.1088/1742-6596/1951/1/012007.

[64] Andersson, A. S., Kalska, B., Häggström, L., Thomas, J. O. Lithium Extraction/ Insertion in $LiFePO_4$: An X-Ray Diffraction and Mössbauer Spectroscopy Study. *Solid State Ion.*2000, *130* (1), 41–52. https://doi.org/10.1016/S0167-2738(00)00311-8.

[65] Arnold, G., Garche, J., Hemmer, R., Ströbele, S., Vogler, C., Wohlfahrt-Mehrens, M. Fine-Particle Lithium Iron Phosphate $LiFePO_4$ Synthesized by a New Low-Cost Aqueous Precipitation Technique. *J. Power Sources*2003, *119–121*, 247–251. https://doi.org/10.1016/S0378-7753(03)00241-6.

[66] Tang, X.-C., Li, L.-X., Lai, Q.-L., Song, X.-W., Jiang, L.-H. Investigation on Diffusion Behavior of Li^+ in $LiFePO_4$ by Capacity Intermittent Titration Technique (CITT). *Electrochimica Acta*2009, *54* (8), 2329–2334. https://doi.org/10.1016/j.electacta.2008.10.065.

[67] Morgan, D., Ven, A. V. der; Ceder, G. Li Conductivity in Li_xMPO_4 (M = Mn,Fe, Co,Ni) Olivine Materials. *Electrochem. Solid-State Lett.*2003, *7* (2), A30. https://doi.org/10.1149/1.1633511.

[68] Guyomard, D., Tarascon, J. M. Li Metal-Free Rechargeable $LiMn_2O_4$/Carbon Cells: Their Understanding and Optimization. *J. Electrochem. Soc.*1992, *139* (4), 937. https://doi.org/10.1149/1.2069372.

[69] Cheng, F., Wang, H., Zhu, Z., Wang, Y., Zhang, T., Tao, Z., Chen, J. Porous $LiMn_2O_4$ Nanorods with Durable High-Rate Capability for Rechargeable Li-Ion Batteries. *Energy Environ. Sci.*2011, *4* (9), 3668–3675. https://doi.org/10.1039/C1EE01795K.

[70] Amatucci, G., Tarascon, J.-M. Optimization of Insertion Compounds Such as $LiMn_2$ O_4 for Li-Ion Batteries. *J. Electrochem. Soc.*2002, *149* (12), K31. https://doi.org/10.1149/1.1516778.

[71] Thackeray, M. M., de Picciotto, L. A., de Kock, A., Johnson, P. J., Nicholas, V. A., Adendorff, K. T. Spinel Electrodes for Lithium Batteries — A Review. *J. Power Sources*1987, *21* (1), 1–8. https://doi.org/10.1016/0378-7753(87)80071-X.

[72] Angelopoulou, P., Paloukis, F., Słowik, G., Wójcik, G., Avgouropoulos, G. Combustion-Synthesized $Li_xMn_2O_4$-Based Spinel Nanorods as Cathode Materials for Lithium-Ion Batteries. *Chem. Eng. J.*2017, *311*, 191–202. https://doi.org/10.1016/j.cej.2016.11.082.

[73] Han, C.-G., Zhu, C., Saito, G., Akiyama, T. Improved Electrochemical Properties of $LiMn_2O_4$ with the Bi and La Co-Doping for Lithium-Ion Batteries. *RSC Adv.*2015, *5* (89), 73315–73322. https://doi.org/10.1039/C5RA13005K.

[74] Thackeray, M. M., Johnson, P. J., de Picciotto, L. A., Bruce, P. G., Goodenough, J. B. Electrochemical Extraction of Lithium from $LiMn_2O_4$. *Mater. Res. Bull.*1984, *19* (2), 179–187. https://doi.org/10.1016/0025-5408(84)90088-6.

[75] Julien, C. M., Mauger, A., Zaghib, K., Groult, H. Comparative Issues of Cathode Materials for Li-Ion Batteries. *Inorganics*2014, *2* (1), 132–154. https://doi.org/10.3390/inorganics2010132.

[76] Yang, G., Wang, L., Wang, J., Yan, W. Tailoring the Morphology of One-Dimensional Hollow $LiMn_2O_4$ Nanostructures by Single-Spinneret Electrospinning. *Mater. Lett.*2016, *177*, 13–16. https://doi.org/10.1016/j.matlet.2016.04.137.

[77] Thackeray, M. M. Manganese Oxides for Lithium Batteries. *Prog. Solid State Chem.*1997, *25* (1), 1–71. https://doi.org/10.1016/S0079-6786(97)81003-5.

[78] Thackeray, M. M. Structural Considerations of Layered and Spinel Lithiated Oxides for Lithium Ion Batteries. *J. Electrochem. Soc.*1995, *142* (8), 2558. https://doi.org/10.1149/1.2050053.

[79] Şahan, H., Göktepe, H., Patat, Ş., Ülgen, A. The Effect of LBO Coating Method on Electrochemical Performance of $LiMn_2O_4$ Cathode Material. *Solid State Ion.*2008, *178* (35), 1837–1842. https://doi.org/10.1016/j.ssi.2007.11.024.

[80] Chu, P. P., Huang, D. L., Fey, G. T. K. Solid-State 7Li NMR Studies of Inverse Spinel $LiNi_xCo_{1-x}VO_4$ Cathode Materials. *J. Power Sources*2000, *90* (1), 95–102. https://doi.org/10.1016/S0378-7753(00)00454-7.

[81] Fey, G. T. K., Wang, K. S., Yang, S. M. New Inverse Spinel Cathode Materials for Rechargeable Lithium Batteries. *J. Power Sources*1997, *68* (1), 159–165. https://doi.org/10.1016/S0378-7753(96)02626-2.

[82] Guo, W. L., Mai, L. Q., Chen, W., Xu, Q., Zhu, Q. Y. Synthesis, Structure and Electrochemical Performance of Nano-Sized $LiNi_{0.5}Co_{0.5}VO_4$. *J. Mater. Sci. Lett.*2003, *22* (14), 1035–1037. https://doi.org/10.1023/A:1024753728795.

[83] Ben Yahia, M., Vergnet, J., Saubanère, M., Doublet, M.-L. Unified Picture of Anionic Redox in Li/Na-Ion Batteries. *Nat. Mater.*2019, *18* (5), 496–502. https://doi.org/10.1038/s41563-019-0318-3.

[84] Liang, L., Hu, G., Jiang, F., Cao, Y. Electrochemical Behaviours of SiO_2-Coated $LiNi_{0.8}Co_{0.1}Mn_{0.1}O_2$ Cathode Materials by a Novel Modification Method. *J. Alloys Compd.*2016, *657*, 570–581. https://doi.org/10.1016/j.jallcom.2015.10.177.

[85] Kong, F., Liang, C., Wang, L., Zheng, Y., Perananthan, S., Longo, R. C., Ferraris, J. P., Kim, M., Cho, K. Kinetic Stability of Bulk $LiNiO_2$ and Surface Degradation by Oxygen Evolution in $LiNiO_2$-Based Cathode Materials. *Adv. Energy Mater.*2019, *9* (2), 1802586. https://doi.org/10.1002/aenm.201802586.

[86] Park, K., Park, J.-H., Hong, S.-G., Choi, B., Seo, S.-W., Park, J.-H., Min, K. Enhancement in the Electrochemical Performance of Zirconium/Phosphate Bi-Functional Coatings on $LiNi_{0.8}Co_{0.15}Mn_{0.05}O_2$ by the Removal of Li Residuals. *Phys. Chem. Chem. Phys.*2016, *18* (42), 29076–29085. https://doi.org/10.1039/C6CP05286J.

[87] Heo, K., Lee, J.-S., Kim, H.-S., Kim, M.-Y., Jeong, H., Kim, J., Lim, J. Ionic Conductor-$LiNi_{0.8}Co_{0.1}Mn_{0.1}O_2$ Composite Synthesized by Simultaneous Co-Precipitation for Use in Lithium Ion Batteries. *J. Electrochem. Soc.*2018, *165* (13), A2955. https://doi.org/10.1149/2.0201813jes.

[88] Gan, Q., Qin, N., Zhu, Y., Huang, Z., Zhang, F., Gu, S., Xie, J., Zhang, K., Lu, L., Lu, Z. Polyvinylpyrrolidone-Induced Uniform Surface-Conductive Polymer

Coating Endows Ni-Rich $LiNi_{0.8}Co_{0.1}Mn_{0.1}O_2$ with Enhanced Cyclability for Lithium-Ion Batteries. *ACS Appl. Mater. Interfaces*2019, *11* (13), 12594–12604. https://doi.org/10.1021/acsami.9b04050.

[89] Kim, Y. J., Cho, J., Kim, T.-J., Park, B. Suppression of Cobalt Dissolution from the $LiCoO_2$ Cathodes with Various Metal-Oxide Coatings. *J. Electrochem. Soc.*2003, *150* (12), A1723. https://doi.org/10.1149/1.1627347.

[90] Aykol, M., Kim, S., Hegde, V. I., Snydacker, D., Lu, Z., Hao, S., Kirklin, S., Morgan, D., Wolverton, C. High-Throughput Computational Design of Cathode Coatings for Li-Ion Batteries. *Nat. Commun.*2016, *7* (1), 13779. https://doi.org/10.1038/ncomms13779.

[91] Cho, Y., Lee, Y.-S., Park, S.-A., Lee, Y., Cho, J. $LiNi_{0.8}Co_{0.15}Al_{0.05}O_2$ Cathode Materials Prepared by TiO_2 Nanoparticle Coatings on $Ni_{0.8}Co_{0.15}Al_{0.05}(OH)_2$ Precursors. *Electrochimica Acta*2010, *56* (1), 333–339. https://doi.org/10.1016/j.electacta.2010.08.074.

[92] Lee, S.-H., Park, G.-J., Sim, S.-J., Jin, B.-S., Kim, H.-S. Improved Electrochemical Performances of $LiNi_{0.8}Co_{0.1}Mn_{0.1}O_2$ Cathode via SiO_2 Coating. *J. Alloys Compd.*2019, *791*, 193–199. https://doi.org/10.1016/j.jallcom.2019.03.308.

[93] Hildebrand, S., Vollmer, C., Winter, M., Schappacher, F. M. Al_2O_3, SiO_2 and TiO_2 as Coatings for Safer $LiNi_{0.8}Co_{0.15}Al_{0.05}O_2$ Cathodes: Electrochemical Performance and Thermal Analysis by Accelerating Rate Calorimetry. *J. Electrochem. Soc.*2017, *164* (9), A2190. https://doi.org/10.1149/2.0071712jes.

[94] Ito, S., Fujiki, S., Yamada, T., Aihara, Y., Park, Y., Kim, T. Y., Baek, S.-W., Lee, J.-M., Doo, S., Machida, N. A Rocking Chair Type All-Solid-State Lithium Ion Battery Adopting $Li_2O–ZrO_2$ Coated $LiNi_{0.8}Co_{0.15}Al_{0.05}O_2$ and a Sulfide Based Electrolyte. *J. Power Sources*2014, *248*, 943–950. https://doi.org/10.1016/j.jpowsour.2013.10.005.

[95] Lim, S. N., Ahn, W., Yeon, S.-H., Park, S. B. Enhanced Elevated-Temperature Performance of $Li(Ni_{0.8}Co_{0.15}Al_{0.05})O_2$ Electrodes Coated with $Li_2O-2B_2O_3$ Glass. *Electrochimica Acta*2014, *136*, 1–9. https://doi.org/10.1016/j.electacta.2014.05.056.

[96] Gan, Z., Hu, G., Peng, Z., Cao, Y., Tong, H., Du, K. Surface Modification of $LiNi_{0.8}Co_{0.1}Mn_{0.1}O_2$ by WO_3 as a Cathode Material for LIB. *Appl. Surf. Sci.*2019, *481*, 1228–1238. https://doi.org/10.1016/j.apsusc.2019.03.116.

[97] Liu, W., Hu, G., Du, K., Peng, Z., Cao, Y. Surface Coating of $LiNi_{0.8}Co_{0.15}Al_{0.05}O_2$ with $LiCoO_2$ by a Molten Salt Method. *Surf. Coat. Technol.*2013, *216*, 267–272. https://doi.org/10.1016/j.surfcoat.2012.11.057.

[98] He, X., Du, C., Shen, B., Chen, C., Xu, X., Wang, Y., Zuo, P., Ma, Y., Cheng, X., Yin, G. Electronically Conductive Sb-Doped SnO_2 Nanoparticles Coated $LiNi_{0.8}Co_{0.15}Al_{0.05}O_2$ Cathode Material with Enhanced Electrochemical Properties for Li-Ion Batteries. *Electrochimica Acta*2017, *236*, 273–279. https://doi.org/10.1016/j.electacta.2017.03.215.

[99] Lee, S.-H., Yoon, C. S., Amine, K., Sun, Y.-K. Improvement of Long-Term Cycling Performance of $Li[Ni_{0.8}Co_{0.15}Al_{0.05}]O_2$ by AlF_3 Coating. *J. Power Sources*2013, *234*, 201–207. https://doi.org/10.1016/j.jpowsour.2013.01.045.

[100] Xie, J., Sendek, A. D., Cubuk, E. D., Zhang, X., Lu, Z., Gong, Y., Wu, T., Shi, F., Liu, W., Reed, E. J., Cui, Y. Atomic Layer Deposition of Stable $LiAlF_4$ Lithium Ion Conductive Interfacial Layer for Stable Cathode Cycling. *ACS Nano*2017, *11* (7), 7019–7027. https://doi.org/10.1021/acsnano.7b02561.

[101] Xiong, X., Wang, Z., Yin, X., Guo, H., Li, X. A Modified LiF Coating Process to Enhance the Electrochemical Performance Characteristics of $LiNi_{0.8}Co_{0.1}Mn_{0.1}O_2$ Cathode Materials. *Mater. Lett.*2013, *110*, 4–9. https://doi.org/10.1016/j.matlet.2013.07.098.

[102] Liu, W., Tang, X., Qin, M., Li, G., Deng, J., Huang, X. FeF_3-Coated $LiNi_{0.8}Co_{0.15}Al_{0.05}O_2$ Cathode Materials with Improved Electrochemical Properties. *Mater. Lett.*2016, *185*, 96–99. https://doi.org/10.1016/j.matlet.2016.08.112.

[103] Tang, Z.-F., Wu, R., Huang, P.-F., Wang, Q.-S., Chen, C.-H. Improving the Electroche-mical Performance of Ni-Rich Cathode Material $LiNi_{0.815}Co_{0.15}Al_{0.035}O_2$ by Removing the Lithium Residues and Forming Li_3PO_4 Coating Layer. *J. Alloys Compd.*2017, *693*, 1157–1163. https://doi.org/10.1016/j.jallcom.2016.10.099.

[104] Wang, J.-H., Wang, Y., Guo, Y.-Z., Ren, Z.-Y., Liu, C.-W. Effect of Heat-Treatment on the Surface Structure and Electrochemical Behavior of $AlPO_4$-Coated $LiNi_{1/3}Co_{1/3}Mn_{1/3}O_2$ Cathode Materials. *J. Mater. Chem. A*2013, *1* (15), 4879–4884. https://doi.org/10.1039/C3TA00064H.

[105] Liu, X., Li, H., Yoo, E., Ishida, M., Zhou, H. Fabrication of $FePO_4$ Layer Coated $LiNi_{1/3}Co_{1/3}Mn_{1/3}O_2$: Towards High-Performance Cathode Materials for Lithium Ion Batteries. *Electrochimica Acta*2012, *83*, 253–258. https://doi.org/10.1016/j.electacta.2012.07.111.

[106] Weigel, T., Schipper, F., Erickson, E. M., Susai, F. A., Markovsky, B., Aurbach, D. Structural and Electrochemical Aspects of $LiNi_{0.8}Co_{0.1}Mn_{0.1}O_2$ Cathode Materials Doped by Various Cations. *ACS Energy Lett.*2019, *4* (2), 508–516. https://doi.org/10.1021/acsenergylett.8b02302.

[107] Sivaprakash, S., Majumder, S. B. Understanding the Role of Zr^{4+} Cation in Improving the Cycleability of $LiNi_{0.8}Co_{0.15}Zr_{0.05}O_2$ Cathodes for Li Ion Rechargeable Batteries. *J. Alloys Compd.*2009, *479* (1), 561–568. https://doi.org/10.1016/j.jallcom.2008.12.129.

[108] Min, K., Seo, S.-W., Song, Y. Y., Lee, H. S., Cho, E. A First-Principles Study of the Preventive Effects of Al and Mg Doping on the Degradation in $LiNi_{0.8}Co_{0.1}Mn_{0.1}O_2$ Cathode Materials. *Phys. Chem. Chem. Phys.*2017, *19* (3), 1762–1769. https://doi.org/10.1039/C6CP06270A.

[109] Huang, X., Rui, X., Hng, H. H., Yan, Q. Vanadium Pentoxide-Based Cathode Materials for Lithium-Ion Batteries: Morphology Control, Carbon Hybridization, and Cation Doping. *Part. Part. Syst. Charact.*2015, *32* (3), 276–294. https://doi.org/10.1002/ppsc.201400125.

[110] Luo, Y., Lu, T., Zhang, Y., Yan, L., Mao, S. S., Xie, J. Surface-Segregated, High-Voltage Spinel Lithium-Ion Battery Cathode Material $LiNi_{0.5}Mn_{1.5}O_4$ Cathodes by Aluminium Doping with Improved High-Rate Cyclability. *J. Alloys Compd.*2017, *703*, 289–297. https://doi.org/10.1016/j.jallcom.2017.01.248.

[111] Reddy, M. V., Jie, T. W., Jafta, C. J., Ozoemena, K. I., Mathe, M. K., Nair, A. S., Peng, S. S., Idris, M. S., Balakrishna, G., Ezema, F. I., Chowdari, B. V. R. Studies

on Bare and Mg-Doped $LiCoO_2$ as a Cathode Material for Lithium Ion Batteries. *Electrochimica Acta*2014, *128*, 192–197. https://doi.org/10.1016/j.electacta.2013.10.192.

[112] Zong, B., Lang, Y., Yan, S., Deng, Z., Gong, J., Guo, J., Wang, L., Liang, G. Influence of Ti Doping on Microstructure and Electrochemical Performance of $LiNi_{0.5}Mn_{1.5}O_4$ Cathode Material for Lithium-Ion Batteries. *Mater. Today Commun.*2020, *24*, 101003. https://doi.org/10.1016/j.mtcomm.2020.101003.

[113] Wang, X., Feng, Z., Hou, X., Liu, L., He, M., He, X., Huang, J., Wen, Z. Fluorine Doped Carbon Coating of $LiFePO_4$ as a Cathode Material for Lithium-Ion Batteries. *Chem. Eng. J.*2020, *379*, 122371. https://doi.org/10.1016/j.cej.2019.122371.

[114] Molenda, M., Bakierska, M., Majda, D., Świętosławski, M., Dziembaj, R. Structural and Electrochemical Characterization of Sulphur-Doped Lithium Manganese Spinel Cathode Materials for Lithium Ion Batteries. *Solid State Ion.*2015, *272*, 127–132. https://doi.org/10.1016/j.ssi.2015.01.012.

[115] Chen, G., An, J., Meng, Y., Yuan, C., Matthews, B., Dou, F., Shi, L., Zhou, Y., Song, P., Wu, G., Zhang, D. Cation and Anion Co-Doping Synergy to Improve Structural Stability of Li- and Mn-Rich Layered Cathode Materials for Lithium-Ion Batteries. *Nano Energy*2019, *57*, 157–165. https://linkinghub.elsevier.com/retrieve/pii/S2211285518309625.

Chapter 2

Recent Advances on the Functionalization of 4-Quinolone Derivatives Via Transition Metal Catalysis

Prasanjit Ghosh*,
Anirban Mandal
and Sajal Das
Department of Chemistry, University of North Bengal, West Bengal, India

Abstract

Functionalization of 4-quinolone scaffold has attracted a great deal of attention from the synthetic community due its widespread biological activity and pharmacokinetic features. Over the last few decades, a variety of noble transition metal catalysts (mainly Pd, Cu, Rh, Ru, and others) have been fostered, enabling for the regioselective and atom-economical stitching of numerous functional groups on this potent moiety. In this perspective, we intend to provide a broader picture on the various transformations and concomitant annulation events for the divergent synthesis of valuable substituted 4-quinolones.

Keywords: cross-coupling reaction, C-H functionalization, asymmetric C-H functionalization, annulation, oxidative carbonylation

* Corresponding Author's Emil: prasanjitmeister@yahoo.com.

In: Transition Metals - An Overview
Editor: Veronica J. Avila
ISBN: 979-8-88697-646-5
© 2023 Nova Science Publishers, Inc.

1. Introduction

Heterocyclic compounds with nitrogen and oxygen atoms are an incredibly important family of active skeletons in organic chemistry, and they are commonly encountered in chemical research, food science, medicine, and other domains. Among them, 4-quinolone derivatives are ubiquitous in the fields of natural products due to their unique structures and diverse biological activities possessing anti-viral [1] antiplasmodial, [2] anti-inflammatory [3], antioxidant [4], anticancer [5], antimicrobial [6], neuroprotective, [7], anti-HIV [8], antimalarial [9] and as well as skin and tissue-infections [10] Some biologically active 4-quinolone embedded scaffolds are represented in figure 1. Specifically, 4-quinolone are considered as a bioisostere of chromen-4-one moiety and exist in equilibrium with its tautomeric form as 4-hydroxy quinoline. In 1962, Leshler and coworkers isolated the nalidixic acid as by-product during the synthesis of chloroquine [11]. Later on, synthetic expertise was mainly engrossed towards the development of various routes for the succinct synthesis of 4-quinolone moiety. Gould-Jacobs, Conrad-Limpach, Camps cyclization, Suzuki-Miyaura reaction and Niementowski cyclization are the forefront classical approach, involving thermal cyclization under harsh conditions [12]. Several other transition metal-catalyzed and metal-free protocols have been well documented in the literature to access this heterocycle [13]. Great strides have been made in recent years towards the simplification of synthetic manifolds and exclusive site-selective C–H functionalization of the 4-quinolone nucleus by improving the reaction conditions and substrate accessibility. Generally, Quinolin-4(1H)-one nucleus encompasses six distinguishable C–H functionalization positions and, owing to the inherent reactivity of the C3 positions of the quinolone core, has witnessed commendable advances over the years. Whilst remote C-H functionalization at the less activated benzenoid core (C-5 to C-8 positions) remains a challenging task in the presence of free C-2 and C-3 positions.

In this regime, the noble metal complexes of Pd, Cu, Ru, Rh and Ir have been extensively fostered for selective C-H functionalization and subsequent annulation en route to carbon-carbon and carbon-heteroatom bond formation.

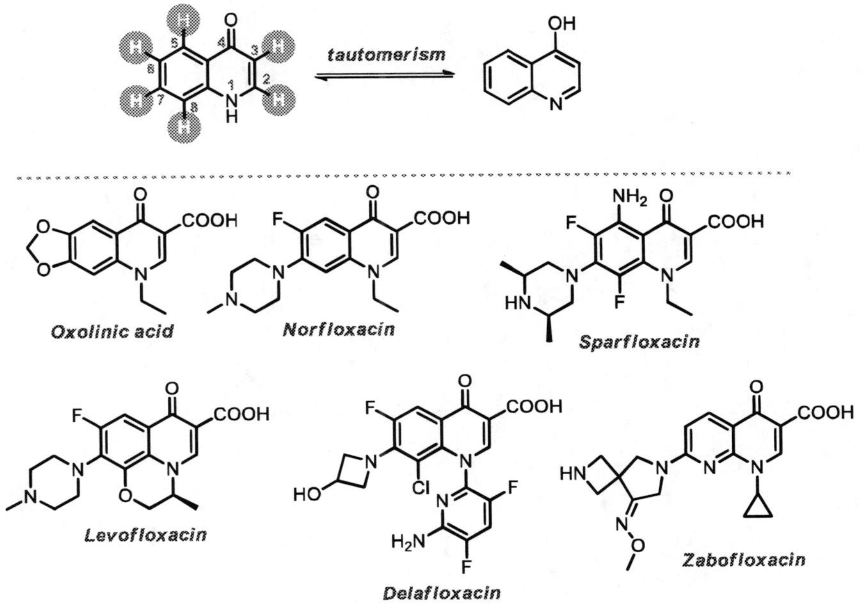

Figure 1. Some of the representative biologically active 4-quinolone scaffolds.

Several reviews have nicely articulated the monumental achievements of these research areas, showcasing the importance, synthesis and medicinal aspects of 4-quinolone derivatives [13, 14]. Herein, we intend to discuss the current state-of-the art of transition metal catalyzed selective C-H and C-C, C-X functionalization and tandem annulation strategies for the streamlined synthesis of polynuclear frameworks. To make the discussion vibrant and confined, special attention is being paid towards (a) substrate scope and limitations, (b) catalytic systems with mechanistic insights and (c) reaction types like arylation, alkynylation, alkenylation, amidation, various cross-coupling reactions (Suzuki, Carbonylative Suzuki, Sonogashira and Heck reactions), amino carbonylations, decarboxylative C-C and C-S cross coupling and –CF2 insertion. Additionally, asymmetric C–H functionalization reaction are nicely elucidated (Figure 2). We anticipate that this intrigue discussion will stimulate the synthetic chemist fraternity to uncover new reactivity paradigms for the construction of C-hetero/N-hetero bond formations in this cutting-edge research arena.

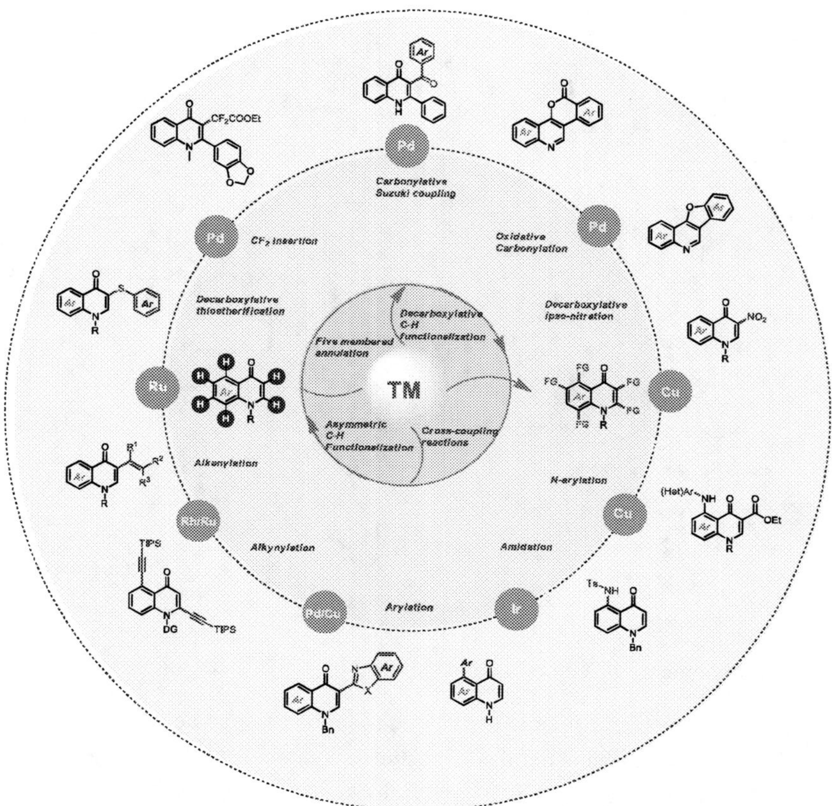

Figure 2. Transition metal catalyzed C-H Functionalization and annulation of 4-quinolone scaffolds.

2. Functionalization

2.1. Cross-Coupling Reactions

2.1.1. Suzuki Coupling and Cyanation

In 2010, Manestch group developed a convenient and divergent route for mono and bis-arylation of 3-iodo and 6-chloro-3-iodo-4(1H)-quinolones under Pd catalysis (Scheme 1) [15]. Particularly, the reaction operated well by the meditation of Pd2(dba)3 as catalyst (4-5 mol %), SPHOS (8-15 mol %) as ligand and K3PO4 as base in the presence of toluene and DMF as solvent. This unified technique tolerated a wide variety of (Het)arylboronic acids, bestowing the anticipated products in reasonable yields. Furthermore, the authors exploited the catalytic activity of both Pd and Copper in cyanation reaction by employing Zn(CN)2/NaCN as a CN source.

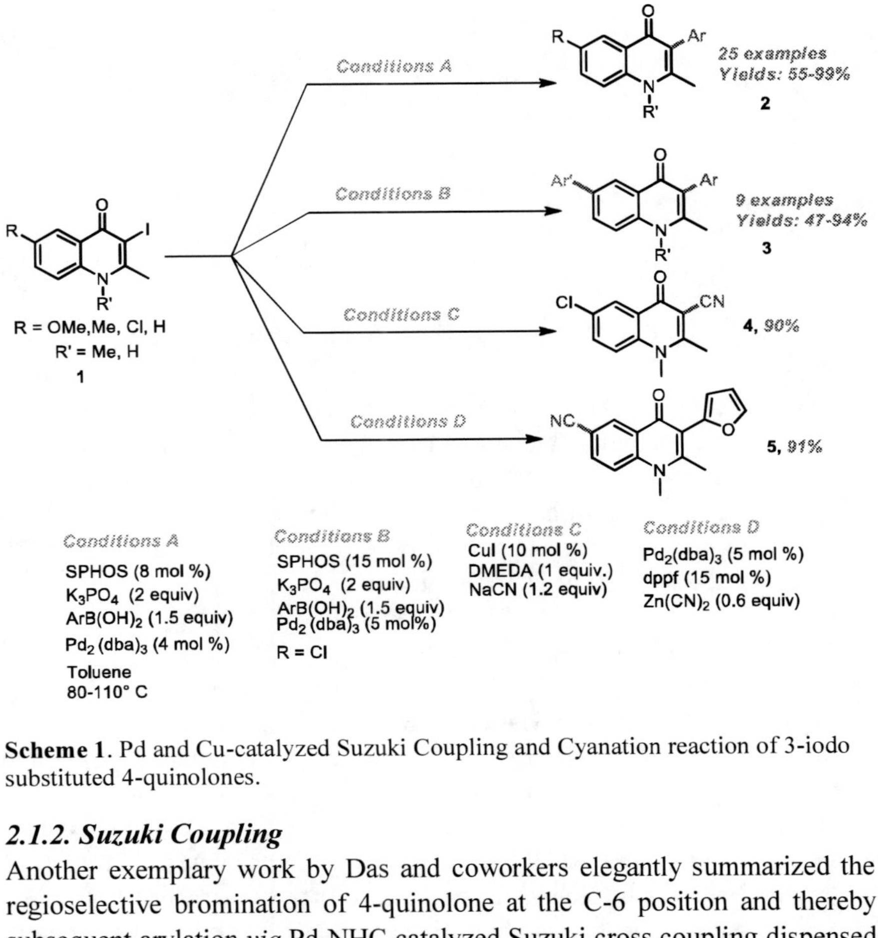

R = OMe,Me, Cl, H
R' = Me, H
1

Conditions A
25 examples
Yields: 55-99%
2

Conditions B
9 examples
Yields: 47-94%
3

Conditions C
4, 90%

Conditions D
5, 91%

Conditions A
SPHOS (8 mol %)
K_3PO_4 (2 equiv)
$ArB(OH)_2$ (1.5 equiv)
$Pd_2(dba)_3$ (4 mol %)
Toluene
80-110° C

Conditions B
SPHOS (15 mol %)
K_3PO_4 (2 equiv)
$ArB(OH)_2$ (1.5 equiv)
$Pd_2(dba)_3$ (5 mol%)
R = Cl

Conditions C
CuI (10 mol %)
DMEDA (1 equiv.)
NaCN (1.2 equiv)

Conditions D
$Pd_2(dba)_3$ (5 mol %)
dppf (15 mol %)
$Zn(CN)_2$ (0.6 equiv)

Scheme 1. Pd and Cu-catalyzed Suzuki Coupling and Cyanation reaction of 3-iodo substituted 4-quinolones.

2.1.2. Suzuki Coupling

Another exemplary work by Das and coworkers elegantly summarized the regioselective bromination of 4-quinolone at the C-6 position and thereby subsequent arylation *via* Pd-NHC catalyzed Suzuki cross coupling dispensed a broad array of 6-(Het)aryl substituted 4-quinolones 8 in good to excellent yields (Scheme 2) [16]. Importantly, the main highlighting feature of the methodology was the formation of desired arylated products within 10 minutes under microwave irradiation.

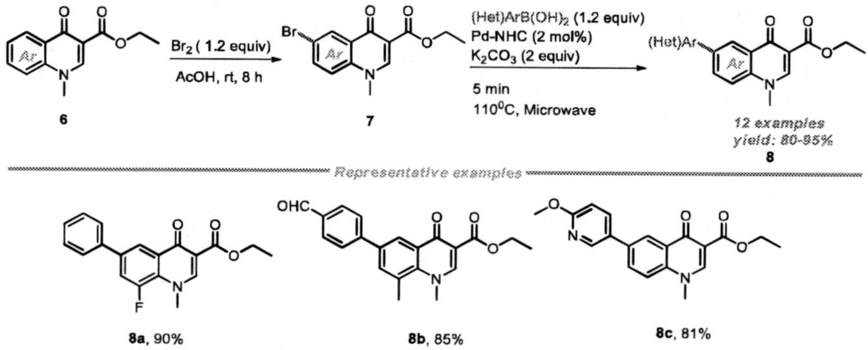

Scheme 2. synthesis of 6-aryl substituted 4-quinolones *via* Pd-NHC catalyzed Suzuki cross-coupling.

Silva et al. unveiled an expeditious route for the synthesis of potent 3-arylquinolin-4(1*H*)-one derivatives 10 under palladium(II)-catalyzed Suzuki-cross coupling (Scheme 3) [17]. Utilization of ohmic heating, ligand free coupling, recyclability up to six times without losing the catalytic activity of the reaction and the use of water as benign reaction medium makes the protocol greener and attractive towards the synthetic community. Electronic effects on boronic acids caused by various substituents have a significant impact on the yield of the intended products.

2.1.3. Carbonylative Suzuki Coupling

Inspired by the aforementioned work, Das et al. laid forward their progress in Pd-NHC catalyzed Carbonylative Suzuki cross coupling of 3-iodo substituted 4-quinolone with various arylboronic acids as coupling unit, witnessing the formation of 2-phenyl-3-aroylquinolin-4(1H)-one derivatives 12 in synthetically useful yields (Scheme 4) [18]. Of note, the authors judiciously engaged Mo(CO)6 as a solid CO surrogate in order to avoid the hazardous CO gas. Mechanistically, the reaction initiated with the generation of active Pd(0) species and participated into oxidative addition with aryl iodides to produce aryl palladium intermediate Int-11a. Following that, in the presence of CO, the aryl palladium intermediate Int-11a is transformed into the acylpalladium species Int-11b, which was created in situ during the decomposition of Mo(CO)6 under the optimum conditions. Later, the complex Int-11b was involved in a transmetalation reaction with arylboronic acid, resulting in the formation of the new complex Int-11c. Finally, the desired product 12 is formed through reductive elimination with the regeneration of Pd(0) species for continuing the next catalytic cycle.

TBAB (0.1 equiv)
Base (1equiv)
$Pd(OAc)_2$ (5 mol%), H_2O
100^0C, ohmic heating
(1.5 equiv)
13 examples
yields: 20-90%
Selected examples
10a, 5 min, 86%
10b, 15 min, 83%
10c, 15 min, 40%
10d, 30 min, 90%

Scheme 3. Pd(II)-catalyzed Suzuki cross coupling under ohmic heating.

2.1.4. Sonogashira Coupling

In 2006, Pal and co-workers utilized tandem Sonogashira coupling–cyclization events under Pd/C-copper catalysis to construct the functionally orchestrated 2-substituted furo[3,2-*c*]quinoline derivatives 15 (Scheme 5) [19]. Under mild circumstances, these reactions produced furo[3,2-*c*]quinolines with high regioselectivity. Regardless of the type of the substituent present at C-2 position of the starting 4-quinolone scaffold, the yield was good. Most importantly, presence of C-2 substituted carboxylate group facilitated the iodide displacement step through intramolecular coordination of carbonyl oxygen to palladium, resulting in higher yields of the polycyclic frameworks.

At the onset, palladium catalyzed activation of the triple bond of the 3-alkynyl quinoline delivered the C-C cross coupling product 13c followed by an intramolecular attack of the oxygen on the activated triple bond with subsequent proton transfer and release of the metal ion to render the requisite product 15. The –NH group of the quinolone ring played a decisive role in the cyclization sequence, and it may have aided the preferred participation of the C-4 quinoline oxygen. Instead of producing the fused cyclic adduct, N-methylated quinolones furnished only the Sonogashira-coupled product 14.

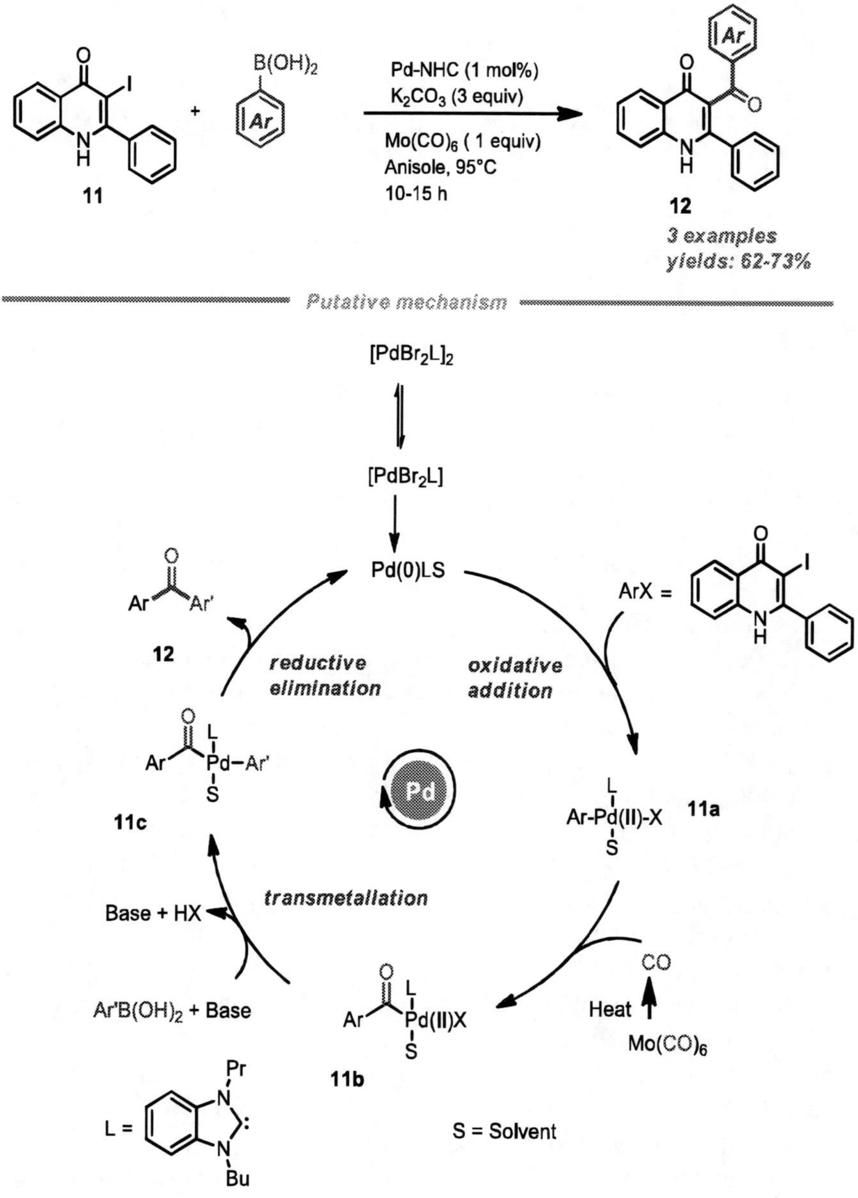

11
$B(OH)_2$
Ar
Pd-NHC (1 mol%)
K_2CO_3 (3 equiv)
$Mo(CO)_6$ (1 equiv)
Anisole, 95°C
10-15 h
12
3 examples
yields: 62-73%

Putative mechanism

$[PdBr_2L]_2$
$[PdBr_2L]$
Pd(0)LS
ArX =
oxidative addition
Ar-Pd(II)-X 11a
CO
Heat
$Mo(CO)_6$
11b
transmetallation
Ar'$B(OH)_2$ + Base
Base + HX
11c
reductive elimination
12
S = Solvent

Scheme 4. Pd-NHC catalyzed carbonylative Suzuki coupling of 3-iodo-4-quinolone derivatives.

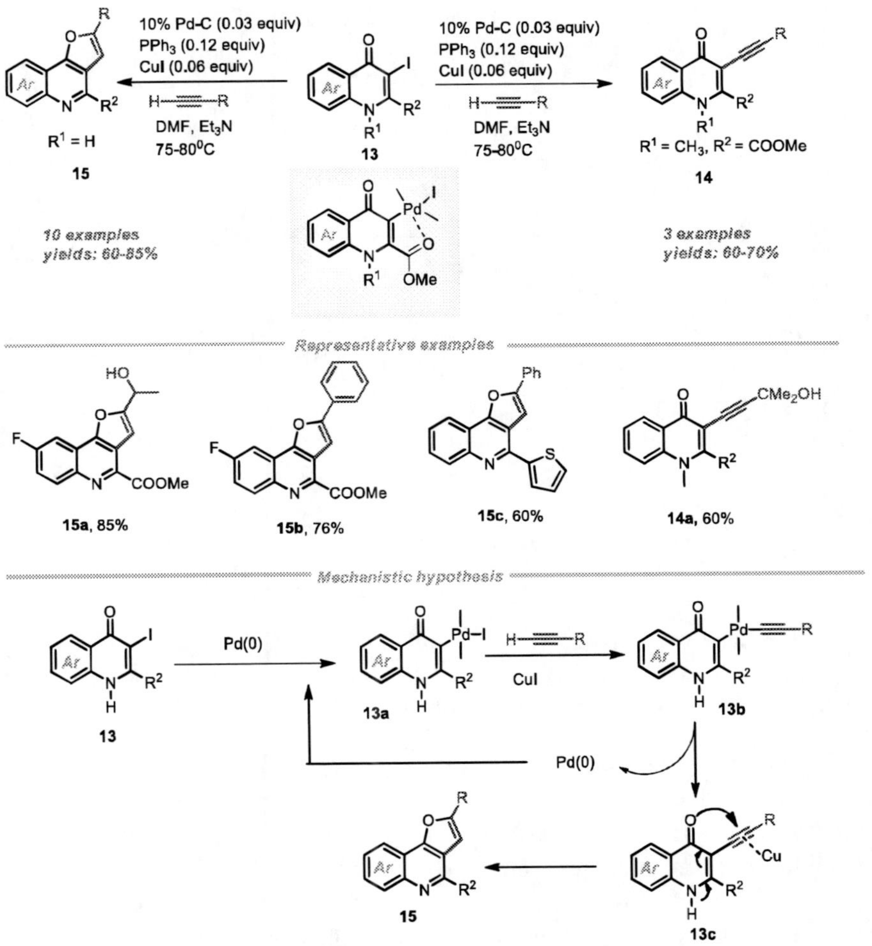

Scheme 5. Synthesis of 2-substituted furo[2,3-*c*]quinolones via Pd/C-copper catalysis.

2.1.5. Suzuki, Sonogashira and Aminocarbonylation

In an ensuing precedent, Corelli and coworkers disclosed the synthesis of 1,3,6-trisubstituted quinolin-4(1*H*)-ones from 1-alkyl-6-bromo-3-iodoquinolin-4(1*H*)-one *via* regioselective sequential Suzuki coupling, Sonogashira coupling and aminocarbonylation process under microwave irradiation (Scheme 6) [20]. This methodology permitted the usage of various arylboronic acids, trimethylsilyl acetylene (TMSA) and amines under the standard reaction conditions, rendering the chemically diverse substituted 4-quinolones 17-20 in decent yields.

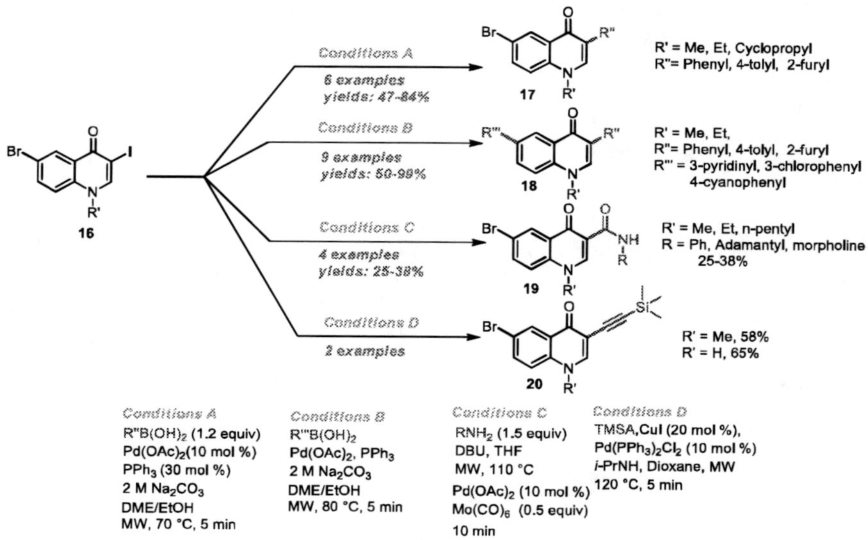

Scheme 6. Palladium catalyzed regioselective sequential Suzuki coupling, Sonogashira coupling and Aminocarbonylation.

2.1.6. Heck Coupling

Because of the sustainability perspective, the synthetic fraternity has always paid close attention to executing organic reactions in an environmentally benign solvent such as water. Silva et al. discovered an unique Pd(II)-catalysed Heck coupling reaction of 3-iodo-1-methyl-quinolin-4(1H)-ones with acrylates as coupling units in this plethora (Scheme 7) [21]. This alkenylation method used 5 mol % Pd(OAc)2 as a catalyst, 1 equiv of K2CO3 as a base, and TBAB as a phase transfer catalyst. The accessibility of the substrate, the short reaction time, the use of a universal solvent (water), the absence of any extra ligand, and the higher yields of the desired products in contrast to classical heating make this approach versatile and promising. The authors also demonstrated the reactivity of α-substituted styrene, yielding the alkenylated product 21f in modest yield.

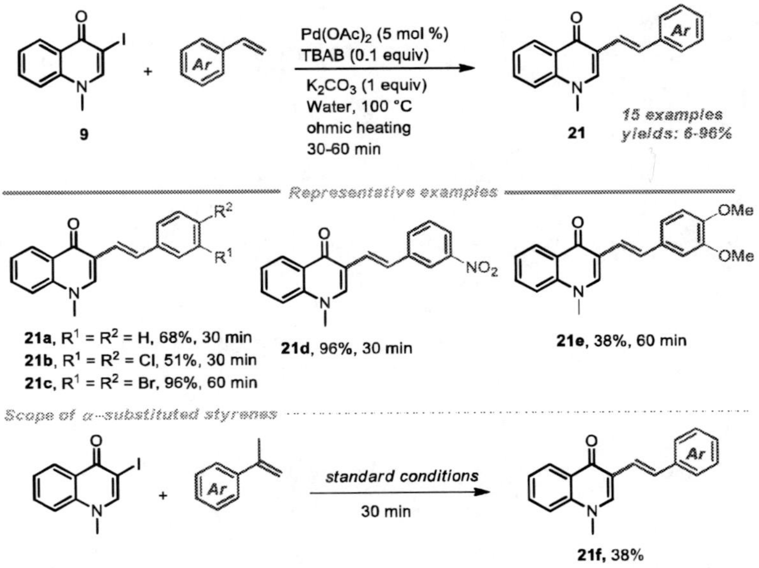

Scheme 7. Pd(II)-catalyzed ohmic heating induced Heck cross coupling of 3-iodo-1-methyl-quinolin-4(1*H*)-ones.

2.1.7. C-N Coupling

In 2011, the first realization of Ullmann type C-N bond forming strategy of 3-halo-4(1*H*)-quinolones with various nucleophiles such as amides, lactams, sulfonamides and NH-containing heterocycles was unravelled by Alami and collaborators (Scheme 8) [22]. The reaction was viable by the cooperative effect of Cu powder/DMEDA (ligand) catalyst system in toluene at 135 °C. To scrutinize the scope and reactivity, the author judiciously engaged electronically diversed indoles, 7-azaindoles, pyrrole, benzimidazole and imidazoles under the standard conditions and afforded a broad spectrum of 3-(*N*-substituted)-4(1*H*)-quinolones 22 in synthetically useful yields.

Apart from the aforementioned primary effort in direct C-N bond production, Das and coworkers revealed a favourable regioselective C-NH2 arylation process of both N-protected and unprotected 4-quinolones under ambient circumstances (Scheme 9) [23]. The method was typically carried out in DCM/EtOH with a stoichiometric quantity of Cu(OAc)2, H2O (as catalyst), and K2CO3 (as base), yielding functionally decorated N-arylated 4-quinolone scaffolds 25 with good yields. A wide range of aryl and (hetero)arylboronic acids worked well in this transition, however 2-thienylboronic acid was ineffective.

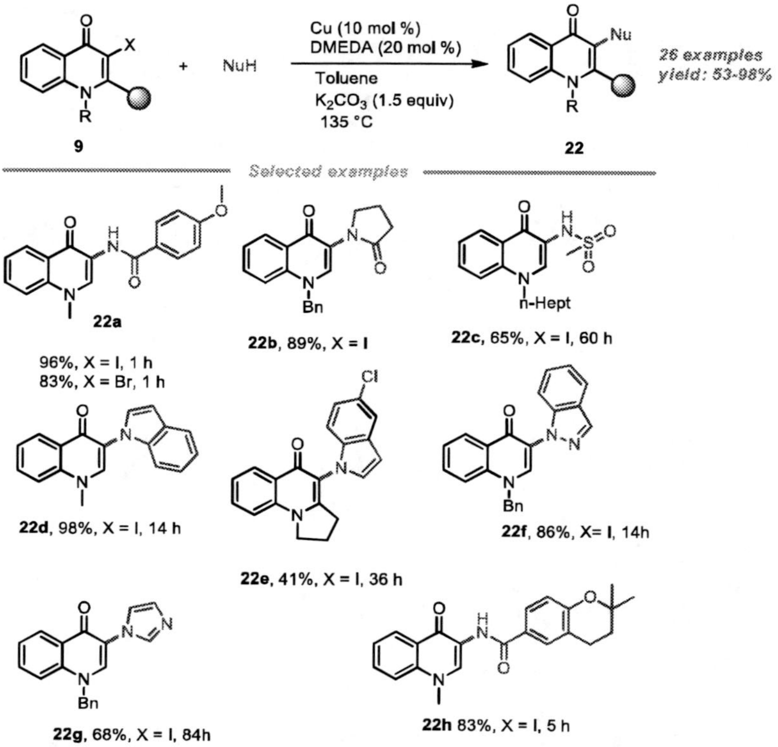

Scheme 8. Cu powder catalyzed C-N cross coupling of 3-halo-4(1*H*)-quinolones.

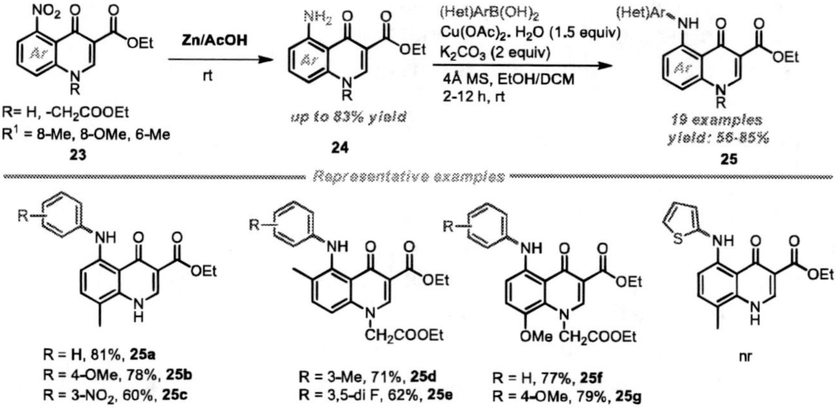

Scheme 9. Cu(II)-catalyzed Chan-Lam type C-N cross coupling of 4-quinolone derivatives.

2.2. Decarboxylative Cross (C-C) Coupling

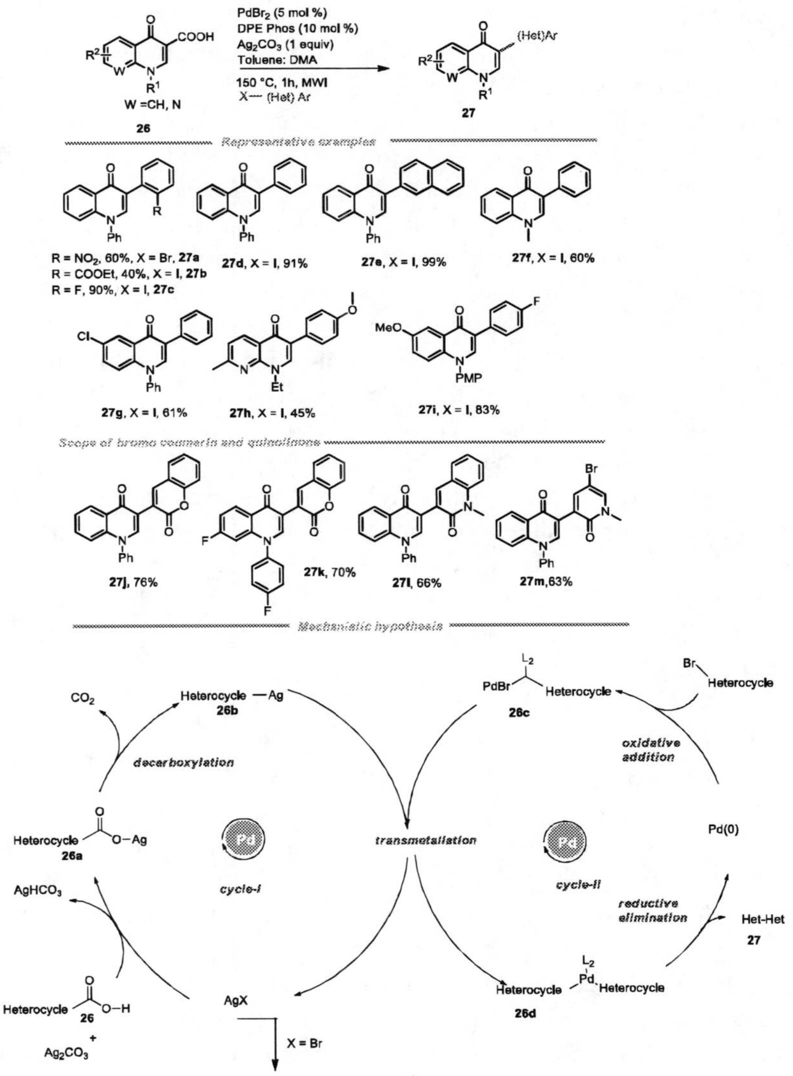

Scheme 10. Pd(II)-catalyzed decarboxylative cross-coupling of 4-quinolone carboxylic acids with heterocyclic halides.

In 2012, Alami and colleagues divulged a potent decarboxylative arylation protocol of quinolin-4(1*H*)-one 3-carboxylic acids with a variety of hetero(aryl) halides, forging an array of 3-hetero(aryl)-4-quinolinones scaffolds (Scheme 10) [24]. Under microwave irradiation, the reaction needed a bimetallic system of $PdBr_2$ and Ag_2CO_3, as well as DPEPhos as a suitable bidentate phosphine ligand. This transformation's reactivity profile has been well proven with several 4-quinolones and electrically diversed aryl iodides and bromides. However, the more challenging aryl chloride could not be used as an accessible coupling unit. The steric and electronic variables caused by the substituents have little effect on the reaction results.

In a subsequent report, the same researchers demonstrated the construction of polynuclear heterocyclic frameworks *via* following similar Pd(II)-catalyzed decarboxylative cross-coupling approach (Scheme 10) [25]. In this catalytic manifold, various quinolone-4-ones 3-carboxylic acid smoothly coupled with a library of 3-bromocoumarins and 3-bromoquinolin-2-ones. A presumptive catalytic cycle is depicted in scheme 10. The reaction began with the extrusion of CO_2 from silver carboxylate intermediate 26a which was produced by the combination of heterocyclic carboxylic acid and silver carbonate. The intermediate 26b then transferred the heterocycle group to a heterocycle palladium(II) complex 26c formed by the oxidative addition of a heterocycle bromide to a palladium catalyst, yielding a bisheterocyclepalladium(II) species 26d. Finally, with the creation of bis-heterocycles 27, the reductive elimination step of Pd(0) species was regenerated.

2.3. Decarboxylative *ipso*-Nitration

In 2015, a facile and expeditious route for the direct access of 3-nitro-4-quinolone derivatives via Cu(II)-catalyzed decarboxylative ipso-nitration was successfully accomplished by Saxena et al. (Scheme 11) [26]. Experiment investigations revealed that the critical effect of both lewis acid [Cu(OAc)2.H2O] and a stoichiometric amount of silver nitrate (as nitrating agent) in attaining the anticipated products 28 in good to outstanding yields. Pertinently, this approach bears a large spectrum of N-alkylated 3-carboxy-4-quinolone substrates with varying electronic straits on the arene backbone. To gain insights into the modus operandi, the authors conducted DFT investigations and measured the electron density charge calculations on the C-3, C-5 and C-8 position. This finding supports the idea that there is enough electron density on the C-3 site to permit decarboxylative nitration. It's worth mentioning that the existence of two carbonyl moiety was entirely responsible for forming a chelate complex with a nitrate ion-containing Cu(II) complex and hence decarboxylative nitration.

Cu(OAc)2.H2O (60 mol %)
AgNO3 (1.2 equiv)
water, 100 °C
12-20 h

26 → 28
18 examples
yield: 80-96%

Representative examples

R = Et, 18 h, 87%, **28a**
R = *sec*-butyl, 16 h, 84%, **28b**
R = Me, 12 h, 82%, **28c**
R = *n*-octyl, 18 h, 88%, **28d**
R = heptyl, 20 h, 86%, **28e**

Plausible mechanism

Cu(OAc)2 + AgNO3 → Cu(OAc)(NO3) + CH3COOAg

Scheme 11. Cu(II)-catalyzed decarboxylative *ipso*-nitration of 4-quinolones.

Use of Chitosan supported Cu-nanoparticles, Narula and coworkers improvised the earlier reported strategy and synthesized the similar scaffolds with the aid of NO2BF4 as nitrating agent (Scheme 12) [27]. The authors recognized the imperative roles of both bi-functional groups such as amine and hydroxyl for stabilizing the chitosan supported Cu-NPs and thereby enhancing the reaction rate. Furthermore, the creation of 3-tetrazolyl bioisosteres 31 with significant antibacterial activity showcased the potential use of this transformation.

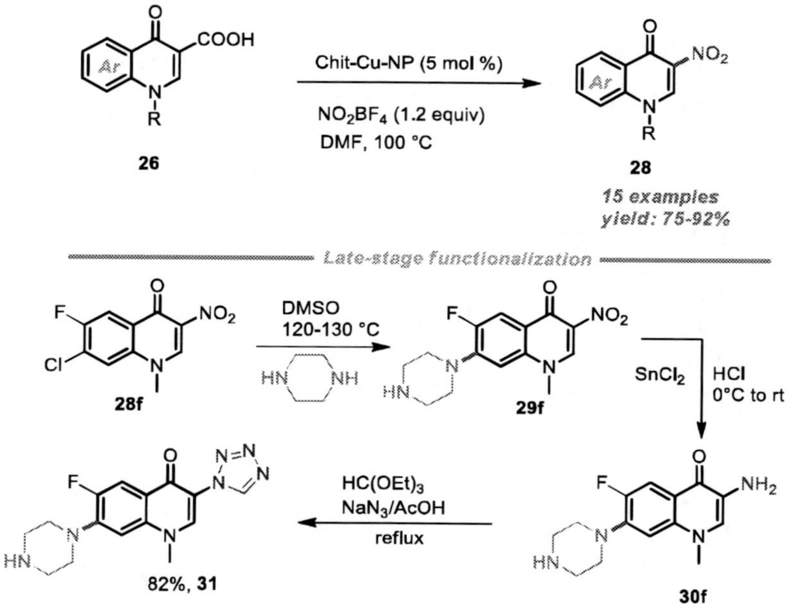

Scheme 12. Chitosan Cu-NPs catalyzed decarboxylative nitration.

2.4. Decarboxylative Thioetherification

In 2016, the first pioneering example of palladium catalyzed intermolecular decarboxylative thioetherification of 4-quinolone scaffolds came from the laboratory of Prof. Zhang (Scheme 13) [28]. The catalytic system of widely accessible $Pd(OAc)_2$ as catalyst, as well as Ag_2CO_3 and PPh_3 as suitable oxidant and ligand, proved to be excellent for the direct –SAr group installation at the C-3 position of 4-quinolone in an air atmosphere. The palette of electron-rich to electron deficient disulfide coupling partners employed for this transformation was really impressive, while alkyl congeners attenuated the desired product formation. Importantly, the author identified the pivotal role of both *N*-protected alkyl group and halo substitution in the backbone of 4-quinolone to promote the C-S bond formation. In Scheme 13, a feasible catalytic cycle is depicted. Organometallic Int-32a was first produced through decarboxylation in the presence of Ag_2CO_3. $Pd(OAc)_2$ then interacted with disulfide to provide the Int-32b. Following that, aryl Pd(II) species Int-32c were produced via a transmetallation reaction between organometallic and Pd(II) intermediate. The intended C-S coupled product 33, as well as Pd(0) species, are formed in the last step by following the reductive elimination pathway. In the presence of air, oxidation of Pd(0) produced Pd(II) species, which could be used to continue the catalytic cycle.

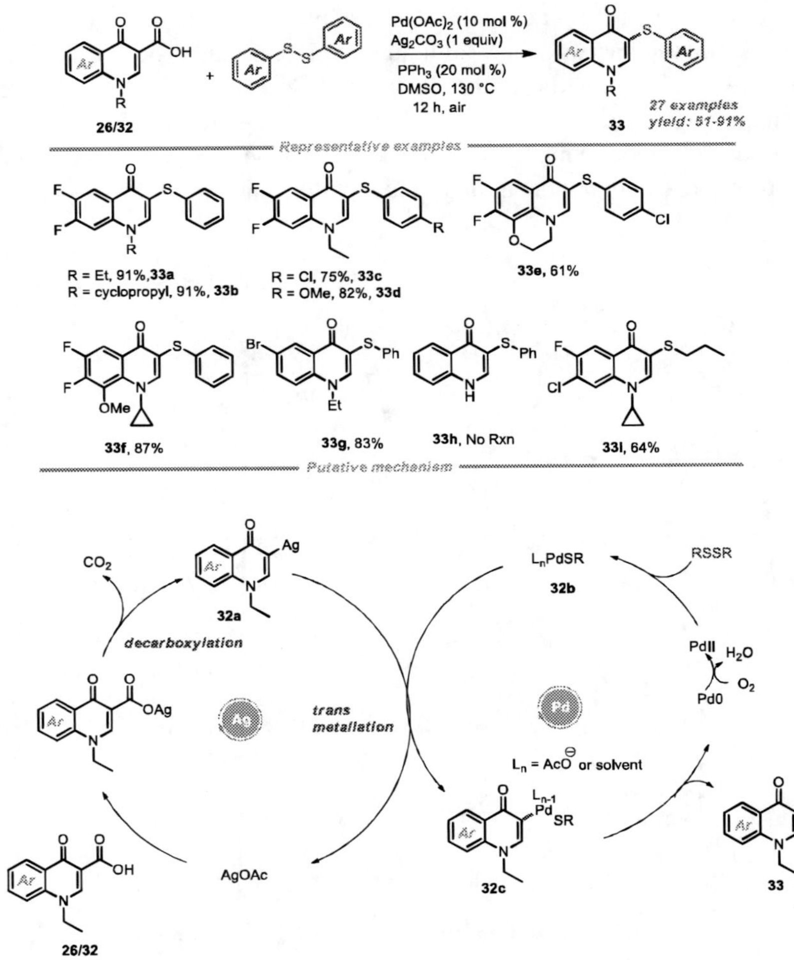

Scheme 13. Pd(II)-catalyzed decarboxylative thioetherification of 4-quinolone derivatives.

2.5. C-H Functionalization

2.5.1. Alkenylation

A fascinating and unprecedented protocol, based on palladium catalysis, has been developed by Ge and co-workers to access the structurally fabricated C-3 alkenylated derivatives of quinolin-4(1*H*)-ones 35-36 (Scheme 14) [29]. Under the standard conditions (1 mol % of $PdCl_2$ as catalyst, 10 mol % $Cu(OAc)_2$ as oxidant under O_2 atmosphere), this method permitted a wide range of terminal acrylates and *N,N*-dimethylacrylamide units to be engaged with good reaction efficiency, although steric hindrance slowed the reaction pace. Several parent molecules with both electron-releasing and electron-withdrawing substituents were able to give the desired products in good to exceptional yields, which was a pleasant surprise. It's worth noting that an *N*-protected 4-quinolone was a pre-requisite for this catalytic method.

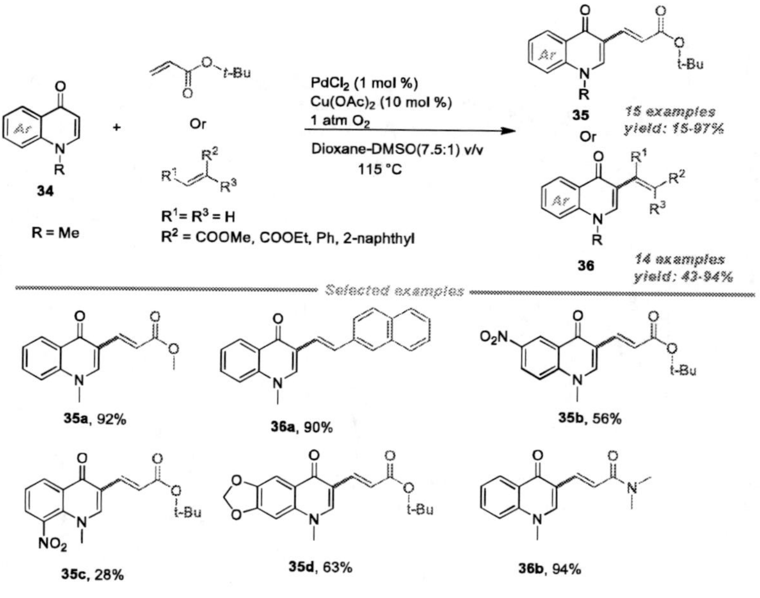

Scheme 14. Pd(II)-catalyzed C-3 alkenylation of 4-quinolone derivatives.

In 2015, Kwon and colleagues delineated 2-pyrimidyl group-aided site-selective C-2 alkenylation and arylation of 4-quinolone derivatives via decarbonylative cross-coupling methodology (Scheme 15) [30]. The utilization of a [Rh(CO)2Cl]2 /(t-BuCO)2O catalytic system was critical in this intriguing report to accelerate the rate of the reaction. The reaction scope was initially evaluated with various alkyl group substituted and α-substituted acrylic acids, yielding alkenylated scaffolds 39 in respectable yields with high E-stereoselectivity. Regardless of the electrical and steric properties of various 4-quinolones and cinnamic acids, all demonstrated to be competent coupling partners. The author also showcased the protocol's adaptability and robustness by selective C-2 arylation and up-scale synthesis. In Scheme 15, a mechanistic rationale for the decarbonylative cross-coupling coupling strategy is outlined. Initially, a pyrimidyl directing group placed the Rh(I) into the C2–H bond of the parent molecule, disbursing a rhodacycle Int-37a . Following that, a suitable anhydride Int-38a created by the equilibrium of the anhydride and acid participated in the oxidative addition with the substrate. Then, it produced the acyl rhodium species Int-37b. Later, complex Int-37c was created by removing carbon monoxide. Finally, reductive elimination proceeded, resulting in the desired product 39 with the concomitant generation of active Rh(I) catalyst.

At the outset of 2015, Patel and co-workers disclosed the first inimitable and convenient route for the straightforward access of pyrano[2,3,4-de]-quinoline derivatives 41 via site-selective ruthenium(II)-catalyzed C-H bond activation followed by an intramolecular cascade annulation of unmasked 2-aryl-quinolin-4(1H)-ones using inexpensive diaryl/alkyl acetylenes (Scheme 16) [31]. In this weaker carbonyl group directed C-H/O-H annulation protocol, the author addressed the necessity of 2 mol % [Ru(p-cymene)Cl2]2 as catalyst and Cu(OAc)2 as suitable oxidant in DCE (as solvent). This transformation proceeded in good to high yields and excellent site-selectivity with structural variations on both quinolone backbone and symmetrical alkyne segment. Nonetheless, aliphatic internal alkynes offered the moderate reactivity. Mechanistic studies through intermolecular competitive experiments between electronically varied two different alkynes bolstered that electro deficient alkyne units were better annulating partners. A catalytic scenario is sketched in (Scheme 16). Initially, an active ruthenium species is generated with the aid of Cu(OAc)2 via successive ligand exchange reaction. Subsequently, aromatization of 2-aryl-quinolin-4(1H)-ones furnished 4-hydroxy-2-arylquinoline derivative under the optimized conditions and it reacted with active Ru-species, offering a five membered ruthenacycle intermediate Int-40b. Then, migratory insertion of alkyne occured into the Ru-carbon bond of Int-40b, resulting in the Int-40c. For unsymmetrical alkynes, the alkyne carbon having higher electron density favoured the insertion into the Ru-carbon bond which accounted for the regioselectivity. Finally, reductive elimination resulted in the expected annulated product 41 with the rejuvenation of catalyst Ru(0). This Ru(0) was then oxidized to an active Ru(II)-catalyst with the aid of oxidant Cu(OAc)2 or through aerial oxidation.

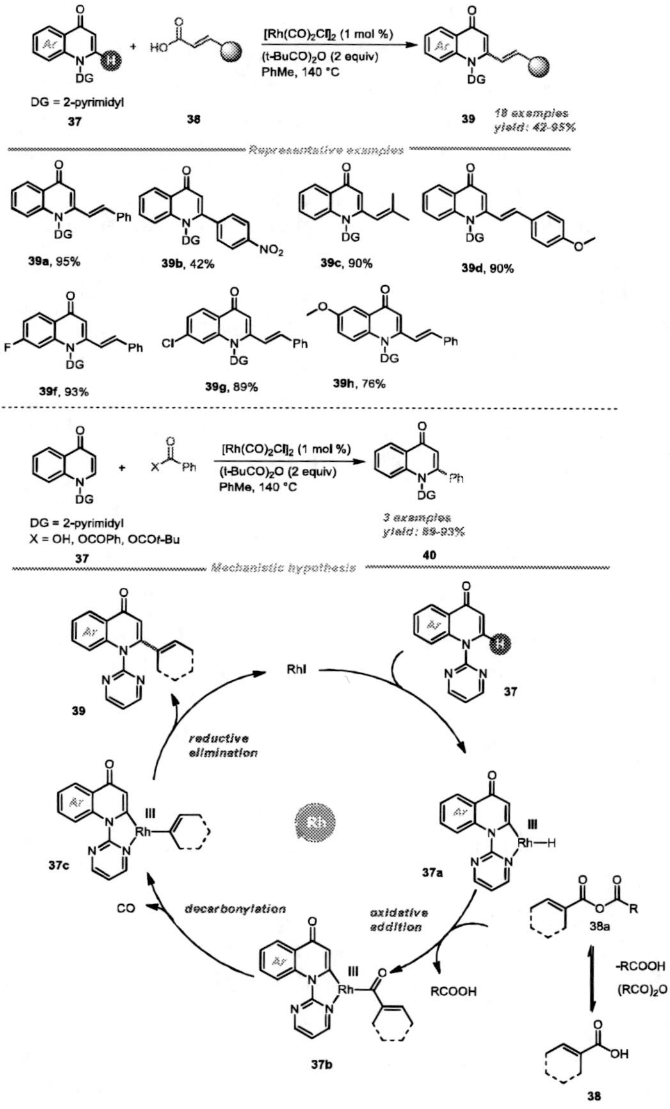

Scheme 15. Rh(I)-catalyzed site selective C-2 alkenylation and arylation of 4-quinolones.

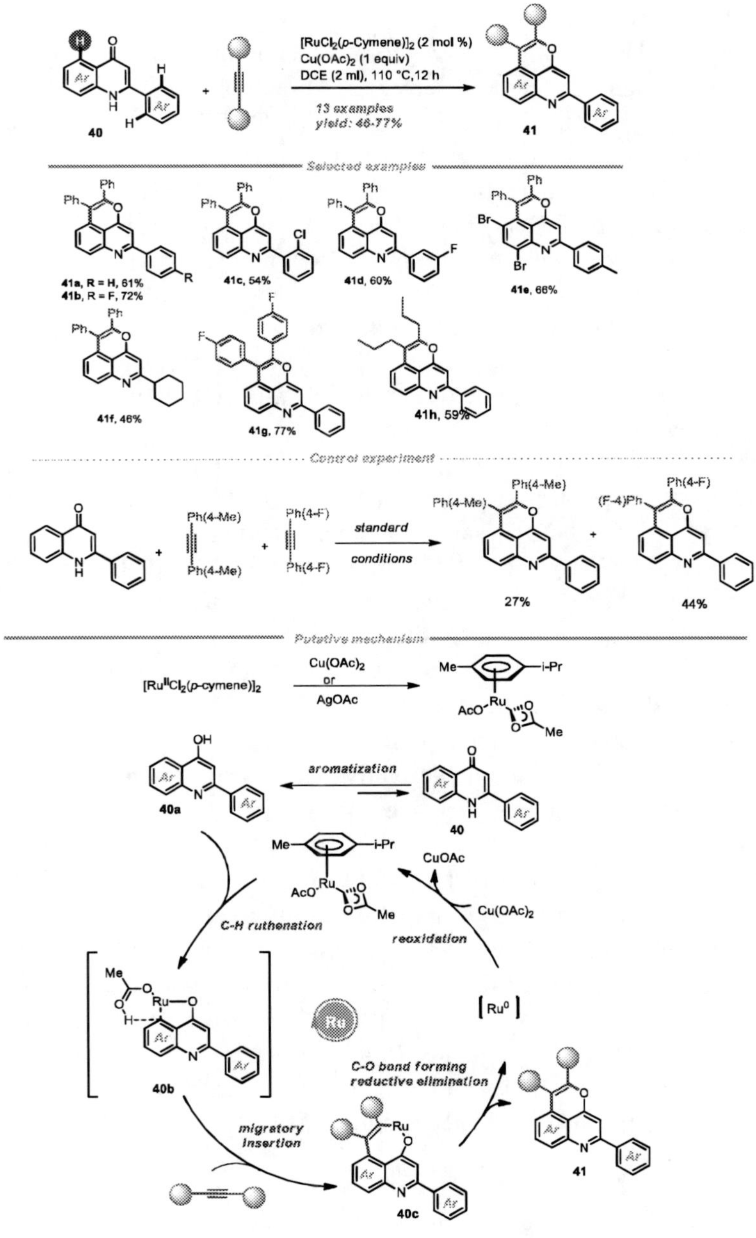

Scheme 16. Ru(II)-catalyzed C-H/O-H annulation of 4-quinolone and internal alkyne.

2.5.2. Alkynylation

In an exemplary work, Hong and co-workers unravelled a compelling Rh(III) and Ru(II)-catalyzed regioselective C-H bond alkynylation of 4-quinolones using an ethynylbenziodoxolone reagent (TIPS-EBX) as a coupling alkyne candidate to prepare diverse C-5 and C-2 alkynylated derivatives 43/44 in good to excellent yields (Scheme 17). [32] Utilization of 5 mol% [Cp*Rh(MeCN)3(SbF6)2]2 in xylene at 80 °C was identified as optimal to achieve a fine conversion of the C-5 monoalkynylated product. An investigation on the substrate scope revealed that N-Benzyl 4-quinolones bearing bromo, iodo, ester and imide groups at the C-3 position functioned well to furnish the corresponding C-5 alkynylated derivatives. Interestingly, the size and electronic properties of the substituents had no significant impact on the reaction efficiency. Of note, C-2 alkynylation also ensued smoothly with the aid of 2-pyrimidyl as DG (directing group) and showed broad functional group tolerance. The author further expanded the theme to a new level by executing both C-2 and C-5 alkynylation in one-pot fashion, affording the dialkynlated derivatives 45 in 63% yield. Remarkably, the ability to synthesize an aaptaminoid derivative 46 in a satisfactory yield indicates the current manifestation's adaptability.

2.5.3. Amidation

In 2018, Samanta group reported the first example of regio-specific C-5 mono sulfamidation of 4-quinolone derivative employing Ir catalysis and sulfonyl azide as amidating source (Scheme 18) [33]. A plausible mechanism is also depicted in Scheme 18. Initially, a putative cationic intermediate [IrCp*(SbF6)2] was formed by ligand exchange reaction with the aid of a silver salt (AgSbF6). Then, it participated in coordination with the carbonyl oxygen atom of 4-quinolone to furnish a cyclometalated intermediate iridacycle Int-47a. Afterwards, tosyl azide reacted with complex Int-47a to produce five-membered complex Int-47b. Furthermore, the release of N2 and migratory insertion of complex Int-47b resulted in the formation of six-membered cyclic complex Int-47c. Finally, protodemetalation of Int-47c delivered the monosulfamidated product 48.

37/42 + TIPS-EBX
TIPS-EBX

$[Cp^*Rh(MeCN)_3(SbF_6)_2]$ (5 mol %)
xylene
80 °C, 12 h
under air

43, R = -Bn, -Me, -Et, cyclopropyl
17 examples
yield: 50-84%

$[Ru(p\text{-cymene})Cl_2]_2$ (4 mol%)
$Zn(OTf)_2$ (16 mol %)
xylene, 60 °C
under air

44, R = Pyrimidyl
12 examples
yield: 47-93%

Selected examples

Scope of C-5 alkynylation

43a, 74% **43b**, 69% **43c**, 47% **43d**, 81% **43e**, 85%

Scope of C-5 alkynylation

44a, 93% **44b**, 75% **44c**, 67%

R = Br, 47%, **44d**
R = OMe, 77%, **44e**
R = CF_3, 55%, **44f**

One-pot C-2/C-5 alkynylation

DG = 2-pyrimidyl

+ TIPS-EBX (2.1 equiv)

$[Cp^*Rh(MeCN)_3(SbF_6)_2]$ (10 mol %)
xylene, 80 °C
63%

45

Late stage functionalization

standard conditions
84%

conc-HCl
80 °C, 73%

aaptaminoid
46

Scheme 17. Rh(III) and Ru(II)-catalyzed C-5/C-2 alkynylation of 4-quinolones.

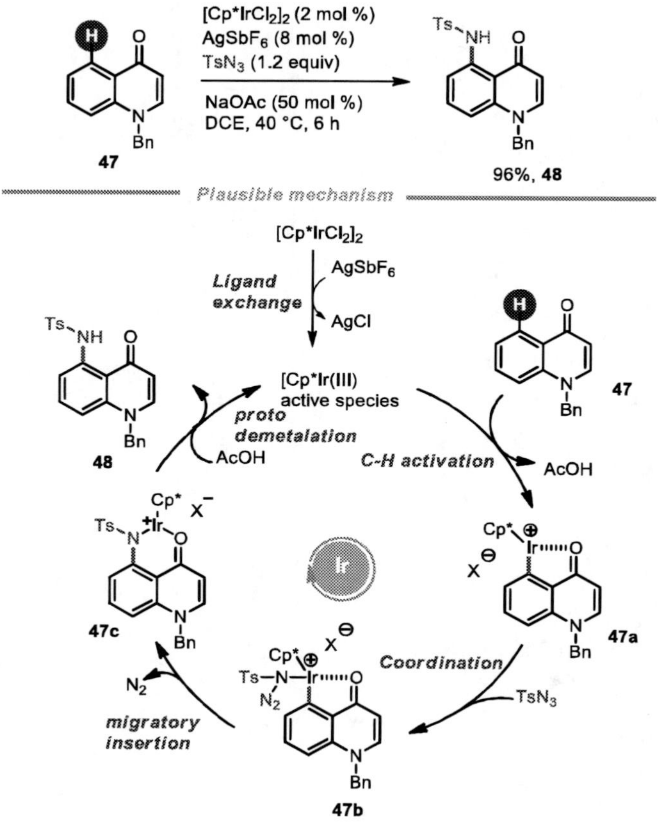

Scheme 18. Ir-catalyzed C-5 amidation of 4-quinolone.

2.5.4. Arylation

Hong and groups were the first to conceptualize that an intriguing CuI/LiOtBu system can be propitious for the succinct synthesis of 3-heteroaryl substituted 4-quinolone derivatives 50 via C-H bond functionalization (Scheme 19) [34]. It was hypothesized that base-assisted cupration of azole produced the Int-49a, which then underwent oxidative addition with N-protected 4-quinolones, resulting in a higher oxidation Cu(III) intermediate Int-49b. The C-3 heteroarylated 4-quinolone derivatives 50 were eventually generated via a reductive elimination route. Steric congestion and the electronic effects of various substituents on the quinolone backbone and synthetically relevant azole substrates (benzothiazole, benzoxazole, benzimidazole, and oxadiazole) had only a minimal effect on the reaction outcome.

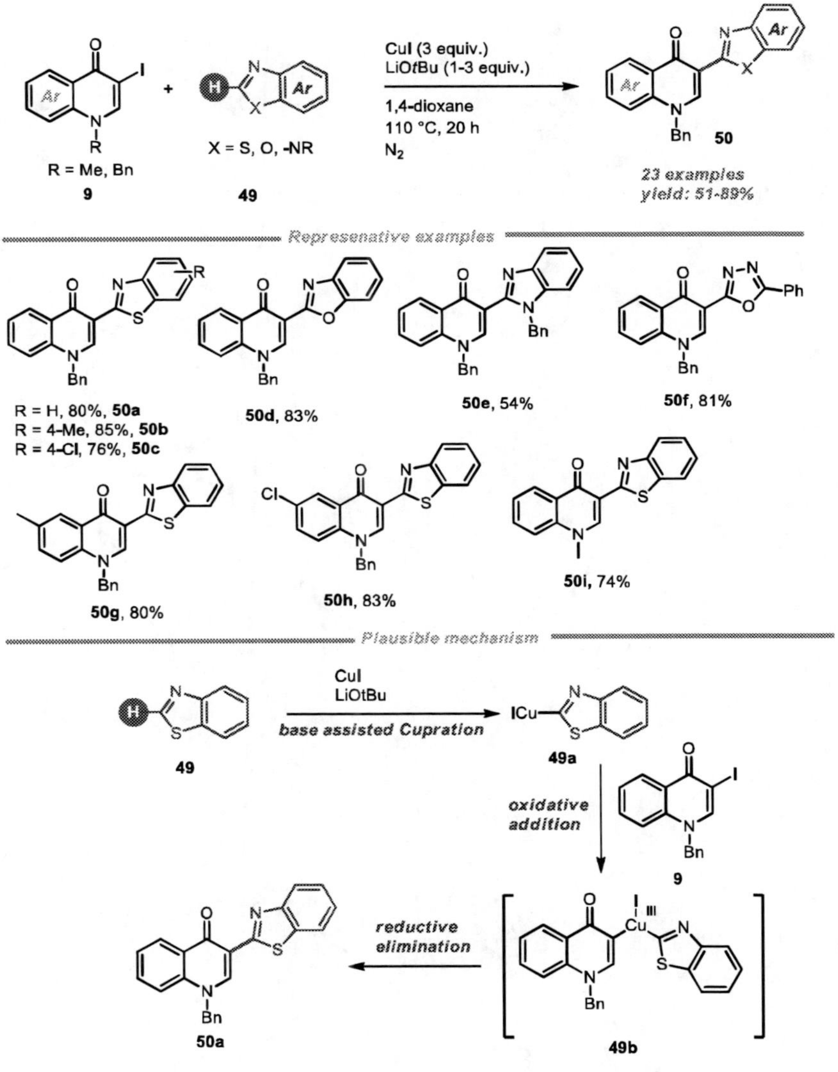

Scheme 19. Cu(I) mediated C3-H heteroarylation of 4-quinolone scaffolds.

One of the first pioneering work of substrate and solvent controlled regioselective C-3/C-5 and C-8 arylation of 4-quinolone scaffolds with diaryliodonium salts as an arylating agent has been unveiled by Kumar and coworkers (Scheme 20) [35]. Detailed investigations revealed that weak

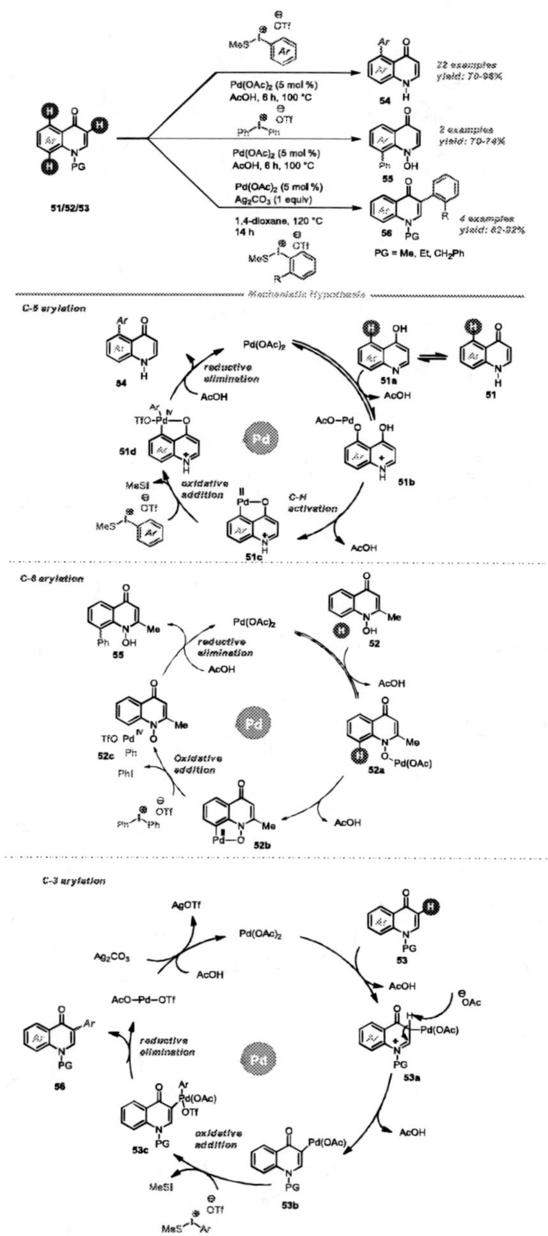

Scheme 20. Pd(II)-catalyzed regioselective C-3/C-5/C-8 arylation of 4-quinolone derivative.

coordinating keto functional group-directed C-5 arylation occurred smoothly in the presence of 5 mol% of $Pd(OAc)_2$ as catalyst in AcOH at 100° C, resulting in synthetically valuable arylated quinolin-4(1*H*)-one building blocks 54-56 with good to outstanding yields. For C-8 arylation, the author employed 1-hydroxy-2-phenylquinolin-4(1*H*)-one as doable coupling partner under the standard reaction conditions and afforded the desired products in high yields. Herein, hydroxyl group played a pivotal role (as DG = directing group) for the regioselective C-8 arylation. Alteration of reaction conditions (use Ag_2CO_3 as an oxidant and dioxane as solvent) allowed in accessing the structurally varied C-3 arylated 4-quinolone units 56 in decent yields. Pleasingly, deuterium scrambling study approved the reversible nature of C-H metalation step (C-5 and C-8) through keto and hydroxyl functionality. A detailed mechanistic pathway is outlined in Scheme 20.

2.6. Miscellaneous Reactions

2.6.1. -CF_2 insertion

Yang et al. employed palladium catalyzed cross-coupling between 3-iodoquinolin-4(1*H*)-ones and ethyl bromodifluoroacetate to successfully install the –CF_2COOEt functionality into the C-3 position of quinolin-4(1*H*)-ones (Scheme 21) [36]. Additionally, the author assessed the method's practicality by structurally altering the natural alkaloid Graveolinine [37] to afford the C-3 ethoxycarbonyldifluoromethylated derivative 59 in synthetically usable yield. Mechanistically, the reaction is believed to be triggered by the reduction of Pd(II) catalyst to an active Pd(0) species in the presence of copper (Scheme 21). It was then participated in oxidative addition with 3-iodo-4-quinolone to generate Int-57a. Meanwhile, the unstable copper ethyl difluoroacetate complex Int-57b was easily produced, and it took part in the transmetalation process with Int-57a to form the intermediate Int-57c. In the final stage, reductive elimination occurred, yielding the desired product 58 via Pd(0) catalyst renewal.

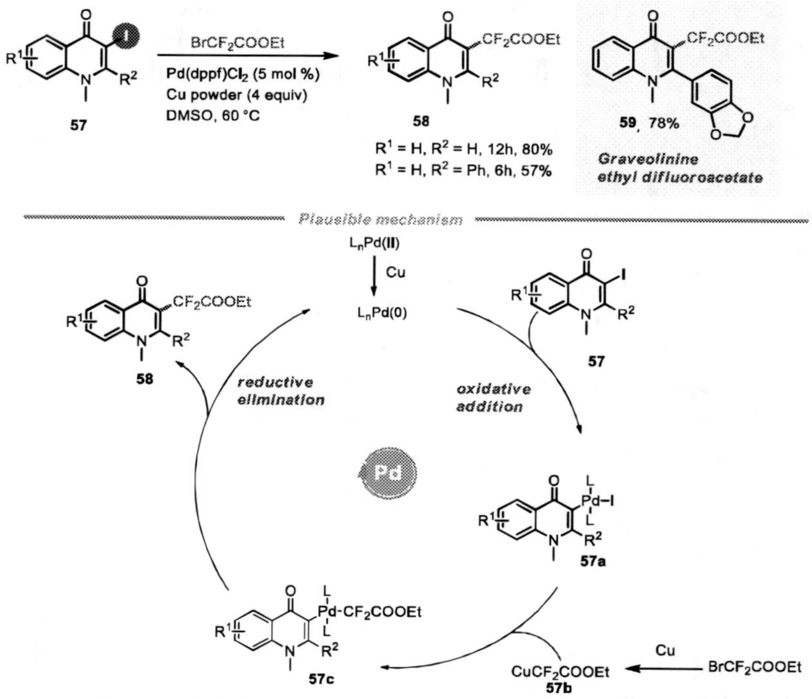

Scheme 21. Synthesis of –CF_2COOEt substituted 4-quinolone derivatives.

2.6.2. Annulation

At the outset of 2018, Peng's group envisioned a one-pot two step synthesis of isoquinolino [2,1-a]quinolinone derivatives 62 in good to excellent yields (Scheme 22) [38]. This one-pot sequential functionalization first involved in a nucleophilic addition of alkynyl bromide to 2-aryl-4-quinolones in a very chemo-, regio- and stereoselective manner, forging the succinct synthesis of alkenylated derivatives and then heating the reaction mixture at elevated temperature in the presence of 5 mol % PdC12 and K2CO3 (as base) allowed the intramolecular C-H alkenylation. The main highlight of this intriguing transformation was the choice of sterically hindered alcohol (tert-pentyl alcohol) as solvent which promoted the nucleophilic attack from N-site of 4-quinolones to alkynyl bromide by forming a hydrogen bond with the carbonyl group. Highly peripherally adorned quinolone with various electronic and steric constrains did not impede the reaction progress. Moreover, para and meta-substituted arylbromo acetylenes were amenable in this cascade annulation protocol whereas ortho-substituents offered inferior outcomes.

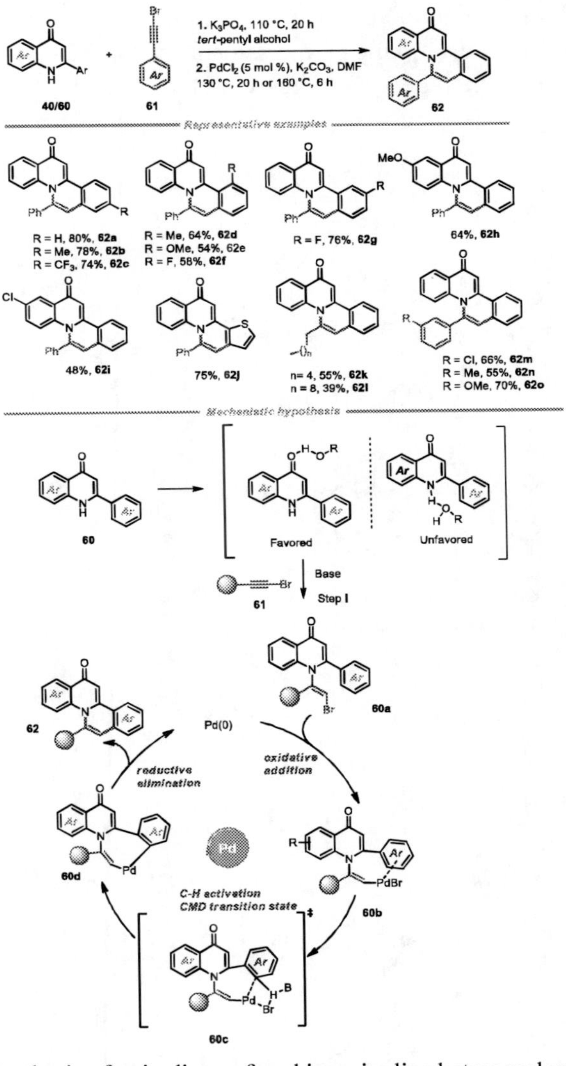

Scheme 22. Synthesis of quinolinone fused isoquinoline heterocycles *via* nucleophilic addition/Pd-catalyzed intramolecular C-H alkenylation.

In an elegant report, Mohan and collaborators divulged a cascade one-pot annulation approach for the synthesis of structurally fabricated phenthridinone heterocycles 64 via successive palladium catalyzed N-benzylation/

intramolecular C-H functionalization of quinolin-4(1H)-one with readily accessible 2-bromobenzyl bromides (Scheme 23) [39]. This operationally simple and step-economic protocol turned out very proficient with electron-deficient substituted coupling partners, dispensing the annulated derivatives in good yields. Surprisingly, 1-bromo-(2-bromomethyl)-naphthalene delivered only N-benzylated product 65 due to steric encumbrance under the standard conditions.

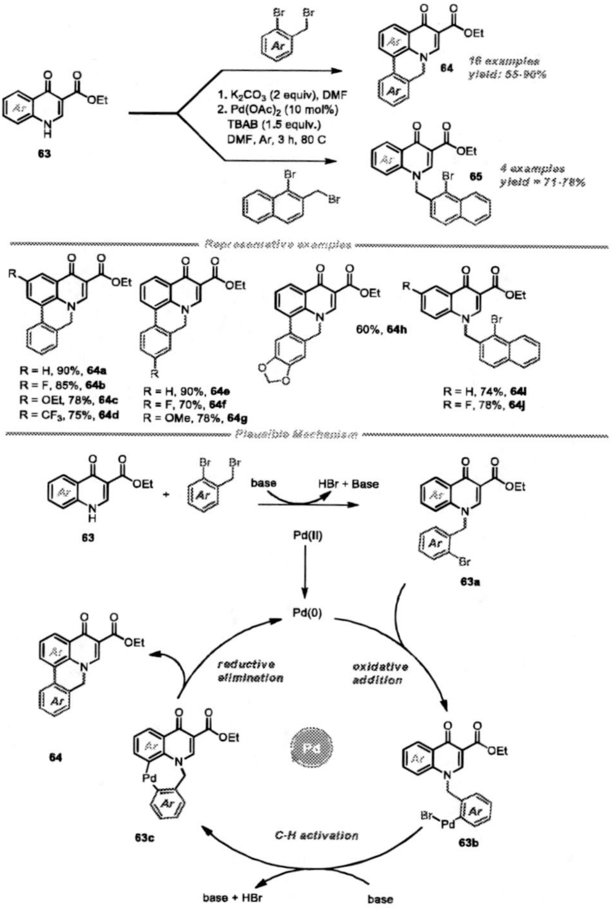

Scheme 23. Pd(II)-catalyzed intramolecular annulation of quinolin-4(1*H*)-ones with 2-bromobenzyl bromide.

2.7. Asymmetric Synthesis

In 2005, Hayashi et al. delineated the first catalytic asymmetric synthesis of 2-aryl-2,3-dihydro-4-quinolones 67 through Rh-catalyzed 1,4-addition of arylzinc reagents with a variety of 4-quinolones (Scheme 24) [40]. The protocol featured mild reaction conditions and exhibited broad functional group tolerance concerning both the coupling partners, and products were isolated in good yield with excellent enantioselectivities (up to 98% *ee*). The author also highlighted the significant role of TMSCl (chlorotrimethylsilane, as additive) and (*R*)-BINAP as ligand on this chiral induction. TMSCl not only facilitates the activation of parent molecule towards 1,4-addition rather it stabilizes the anticipated product by forming a silyl enol ether.

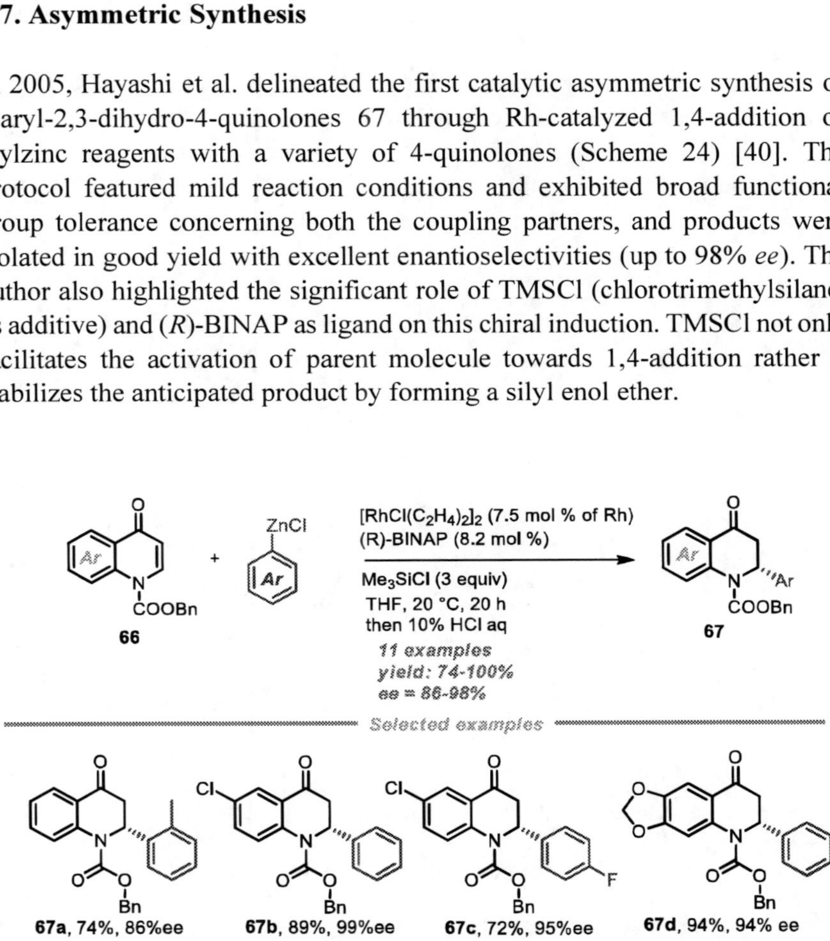

Scheme 24. Rh-catalyzed 1,4-addition of arylzinc reagent to quinolin-4(1*H*)-one derivatives.

In a related study, Harutyunyan and coworker unveiled a propitious strategy for the more challenging enantioselective addition of Grignard reagent towards the less reactive 4-quinolone derivatives at room temperature (Scheme 25) [41]. The chiral diphosphine ligand encapsulated with catalytic amount of CuBr. SMe2 was found to be essential to inducing stereoselectivity in the anticipated products, conferring better steric interactions with the parent molecule. Of note, this transformation endured a wide range of N-protected 4-quinolones, albeit the addition of lewis acid (TMSBr) was necessary for N-Bn and N-Me protected quinolone moieties. Various alkyl and aryl substituted Grignard reagent also proved to be successful under the standard reaction conditions, furnishing the chiral 2-substituted-tetrahydroquinolone products with high enantiomeric excess. Electronic factors of the substituents on the arene backbone of N-CBz-protected 4-quinolone derivative had a substantial effect on the progress of the reaction.

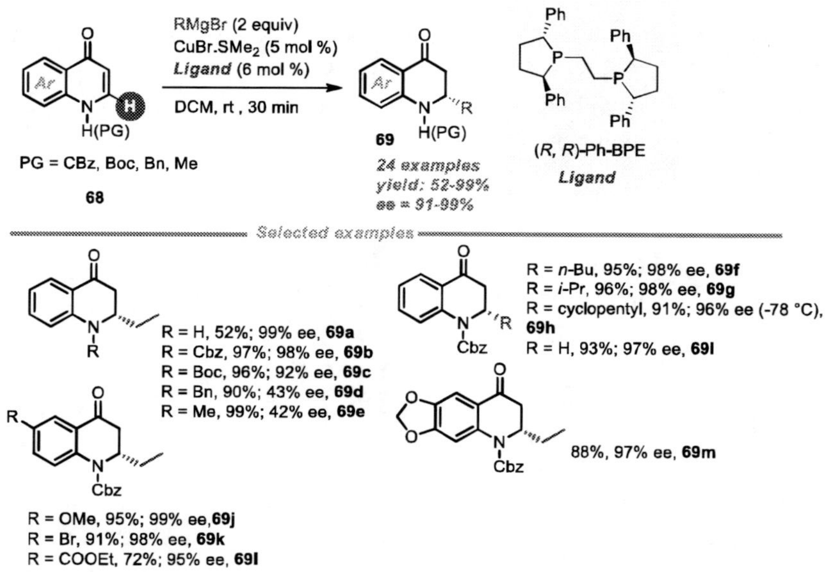

Scheme 25. enantioselective addition by Cu(I)-complex with (*R*,*R*)-Ph-BPE.

Very recently, Phansavath and colleagues unfolded the reactivity of in-house developed Rh(III)-complex for the succinct synthesis of enantiomerically enriched alcohol 71 with high enantioselectivities (Scheme 26) [42]. This asymmetric transfer hydrogenation (ATH) was carried out under mild conditions with an azeotropic mixture of $HCOOH/Et_3N$ (5:2) as hydride source, permitting the tolerance of diversely substituted 4-quinolone scaffolds. Furthermore, the author engaged several *N*-protected 4-quinolone (Pg = Bn, Me and Ms) derivatives under optimal conditions, yielding negative results. Additionally, gram-scale synthesis and late-stage functionalization of the desired chiral alcohol products showcased the utility of the protocol.

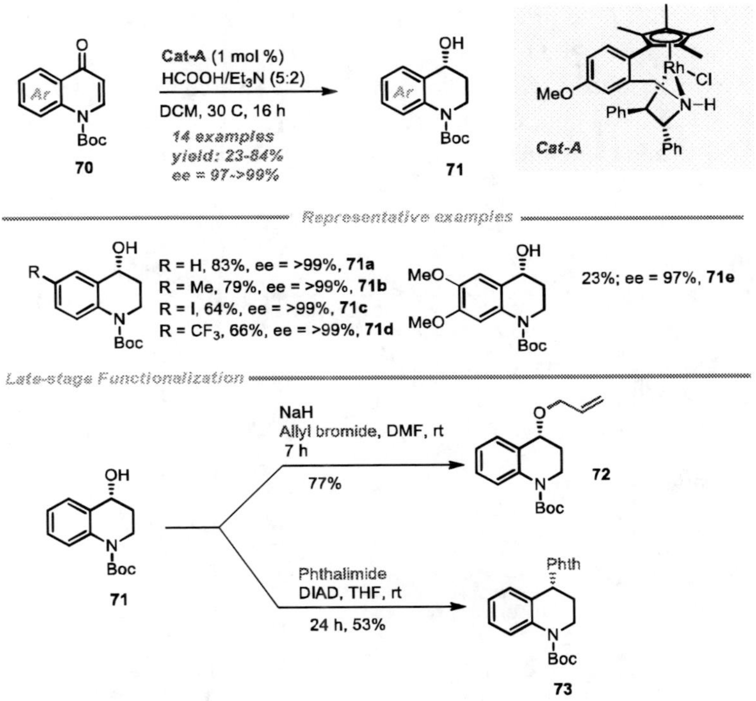

Scheme 26. Rh(III)-catalyzed asymmetric transfer hydrogenation.

2.8. Oxidative Carbonylation

Jiang et al. revealed a propitious strategy for the direct access of polycyclic aromatic hydrocarbons (PAHs) 75-76 under palladium catalysis *via* intramolecular oxidative cyclocarbonylation and oxidative cyclization methodology (Scheme 27) [43]. In this C-H activation/annulation manifold, 10 mol % of $Pd_2(dba)_3$ as catalyst and stoichiometric amount of *p*-TsOH. H_2O as additive, $Cu(OAc)_2$. H_2O as oxidant were utilized, affording the fused frameworks in good to excellent yields. Substrates with strong electron with-drawing and releasing substituents on the arene backbone and C-3 aryl segment did not create any major hindrance during the progress of the reaction. Mechanistically, the C-H carbonylation/annulation initiated with the formation of active $Pd(OTs)_2$ species by the reaction of $Pd_2(dba)_3$ with TsOH. H_2O and $Cu(OAc)_2.H_2O$. Subsequently, C-3 aryl C-H bond activation led to the generation aryl-palladium species Int-74a in the presence of $Pd(OTs)_2$ and reacted with CO to procure Int-74b. It readily transformed into Int-74c with the elimination *p*-TsOH. Lastly, the desired annulated heterocycle 75 was formed *via* reductive elimination process with the concomitant generation of Pd(0)species. Without CO, the catalytic cycle continued with the generation of Int-74d and the reductive elimation route led to the production of PAHs 76.

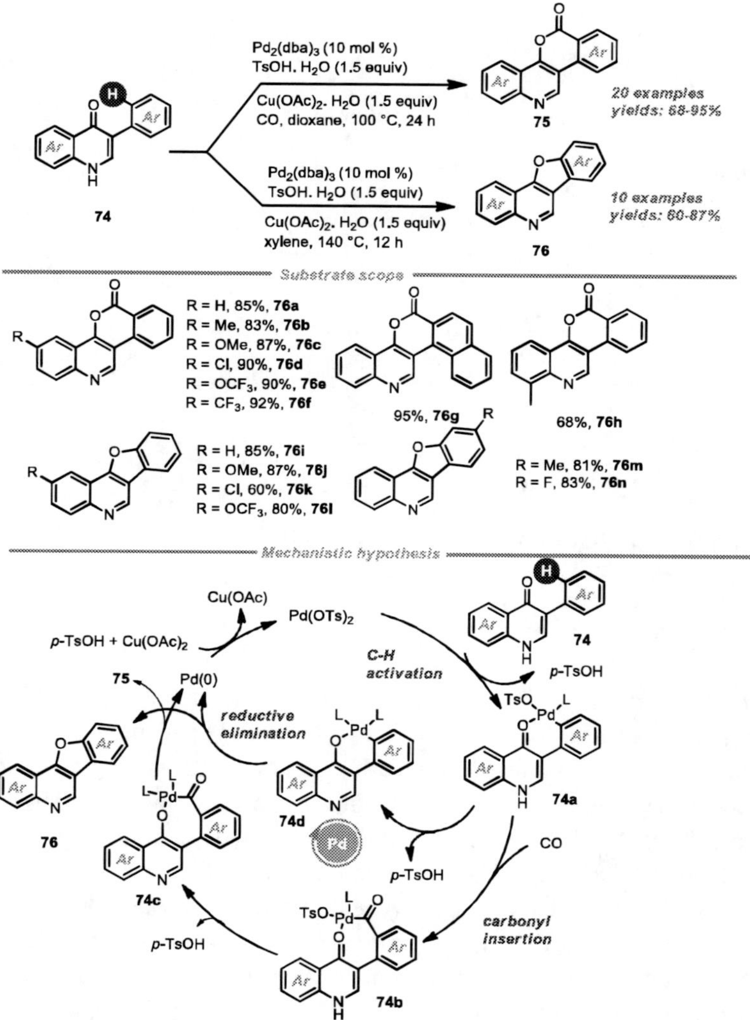

Scheme 27. Palladium catalyzed oxidative cyclocarbonylation and oxidative cyclization of 4-quinolone derivatives.

Conclusion

Considering the omnipresence of 4-quinolone backbone in plentiful organic scaffolds and pharmaceuticals, exemplary efficient synthetic arsenals for C-H and N-H functionalization/annulation of 4-quinolones have been emerged thoroughly. To keep this discussion informatively compelling and concise, we have showcased past developments and current ascendants in the realm of transition-metal catalyzed selective C-H/N-H functionalization and concomitant annulation protocols to fabricate the 4-quinolone nucleus in a step and atom-economical manner. Of note, novel transition-metal catalysts (Pd, Ru and Rh) have been extensively fostered, while the use of earth abundant metals (Co, Ni, Cu and Fe) remains elusive and must be developed for sustainability purposes. Looking closer, enantioselective transformations and target oriented design have not been widely embraced and require major effort from the synthetic community. Despite a number of decisive breakthroughs in various cross-coupling processes and decarboxylative cross-coupling reactions towards the formation of C-C, C-N and C-S bonds, there are still many prospects for C-B, C-P and C-O bonds to be explored. Furthermore, utilization of carbonyl functionality in 4-quinolone scaffold as weak DG (directing group) towards C-H activation modality is still in its infancy and should be a priority for future research. As a consequence, we anticipate that this current reaction compendium will serve as a booster for medicinal and synthetic chemists to surmount the inherent hurdles and make this arena more sparkling and vibrant.

References

[1] Vandekerckhove, S., Desmet, T., Tran, H. G., de Kock, C., Smith, P. J., Chibale, K. and D'hooghe, M. (2014). Synthesis of halogenated 4-quinolones and evaluation of their antiplasmodial activity. *Bioorg. Med. Chem. Lett.* 24: 1214–1217.

[2] Greeff, J., Joubert, J., Malan, S. F. and Van Dyk, S. (2012). Antioxidant properties of 4-quinolones and structurally related flavones. *Bio. Med. Chem., 20*: 809–818.

[3] Mermer, A., Demirbaş, N., Şirin, Y., Uslu, H., Azdemir, Z. and Demirbaş, A. (2018). Conventional and Microwave Prompted Synthesis, Antioxidant, Anticholinesterase Activity Screening and Molecular Docking Studies of New Quinolone-Triazole Hybrids. *Bioorg. Chem.,* 78: 236–248.

[4] Kumar, A., Fernandes, J. and Kumar, P. (2016). Synthesis and Biological Evaluation of Some Novel Pyrazoline incorporated 2-Quinolones. *Res. J. Pharm. Tech.* 9: 2257–2260.

[5] Mohamed, Y. M., Solum, E. J. and Eweas, A. F. (2018). Synthesis, anti-bacterial evaluation, and docking studies of aza-isoflavone analogues generated by palladium catalyzed cross-coupling. *Monatsh. Chem.*, *149*: 1857–1864.

[6] (a) Wube, A., Guzman, J. D., Hufner, A., Hochfellner, C., Blunder, M., Bauer, R., Gibbons, S., Bhakta, S. and Bucar, F. (2012). Synthesis and Antibacterial Evaluation of a New Series of *N*-Alkyl-2-alkynyl/(E)-alkenyl-4-(1*H*)-quinolones. *Molecules,* 17: 8217–8240; (b) Plevova, K., Briestenska, K., Colobert, F., Mistrikova, J., Milata, V. and Leroux, F. R. (2015). Synthesis and biological evaluation of new nucleosides derived from trifluoromethoxy-4-quinolones. *Tetrahedron Lett.,* 56: 5112–5115.

[7] Fukushima, Y., Tsuyuki, Y., Goto, M., Yoshida, H. and Takahashi, T. (2020). Novel Quinolone Nonsusceptible Streptococcus canis Strains with Point Mutations in Quinolone Resistance-determining Regions and Their Related Factors. *Jpn. J. Infect.Dis. 73*: 242–249.

[8] Sales, E. M. and J. D. Figueroa-Villar, (2016). *World J. of Pharm. and Pharmac. Sci.,* 5: 253–268.

[9] (a) Vinayaka, A. C., Sadashiva, M. P., Wu, X., Biryukov, S. S., Stoute, J. A., Rangappa, K. S. and Gowda, D. C. (2014). Facile synthesis of antimalarial 1,2-disubstituted 4-quinolones from 1,3-bisaryl-monothio-1,3- diketones. *Org. Biomol. Chem.,* 12: 8555–8561; (b) Fan, Y. L., Cheng, X. W., Wu, J. B., Liu, M., Zhang, F. Z., Xu, Z. and Feng, L. S. (2018). *Eur. J. Med. Chem.,* 146: 1–14.

[10] (a) Wolfson, J. S. and Hooper, D. C. (1989). Treatment of Genitourinary Tract Infections with Fluoroquinolones: Activity In Vitro, Pharmacokinetics, and Clinical Efficacy in Urinary Tract Infections and Prostatitis. *Antimicrob. Agents Chemother. 33*: 1655–1661. (b) Bambeke, F. V., Michot, J.-M., Eldere, J. V. and Tulkens, P. (2005). Quinolones in 2005: an update. *Clin. Microbiol. Infect.* 2005, 11: 256–280.

[11] Emmerson, A. M. and Jones, A. M. (2003). The Quinolones: decades of development and use. *J. Antimicrob. Chemother.,* 51: 13–20.

[12] (a) Fuson, R. C. and Burness, D. N. (1946). *J. Am. Chem. Soc.,* 68:1270–1272. (b) Joule, J. A. and Mills, K. (2000). *4th ed.; Blackwell Science: Oxford, MA, USA.* 6: 1–589. (c) Boteva, A. A. and Krasnykh, O. P. (2009). *Chem. Heterocycl. Compd.,* 45: 757–785. (c) Shen, C., Wang, A., Xu, J., An, Z., Loh, K. Y., Zhang, P. and X. Liu, (2019). *Chem.*, 5: 1059–1107.

[13] Ghosh, P. and Das, S. (2017). Synthesis and Functionalization of 4-quinolones: A progressing story, 4466-4516.

[14] (a) Singh, G., Devi, V. and Monga, V. (2020) Recent Developments in the Synthetic Strategies of 4-Quinolones and Its Derivatives, *ChemistrySelec*t, 5: 14100-14129; (b) Dhiman, P.; Arora, N.; Thanikachalam, P. V. and Monga, V. (2019). Recent advances in the synthetic and medicinal perspective of quinolones: A review. *Bioorg. Chem.,* 92: 103291.

[15] Cross, R. M. and Manetsch, R. (2010). Divergent Route to Access Structurally Diverse 4-Quinolones via Mono or Sequential Cross-Couplings. *J. Org. Chem.*, 75: 8654-8657.

[16] Gupta, S., Ghosh, P., Dwivedi, S. and Das. S. (2014). Synthesis of 6-aryl substituted 4-quinolones. *RSC Adv.,* 4: 6254-6260.

[17] Pinto, J., Silva, V. L. M., Silva, A. M. G., Santos, L. M. N. B. F., Silva, A. M. S. (2015). Ohmic heating assisted synthesis of 3-arylquinolin-4(1H)-ones by a reusable and ligand-free Suzuki-Miyaura reaction in water. *J. Org. Chem.*, 80: 6649-6659.

[18] Ghosh, P., Ganguly, B. and Das, S. (2017). Pd-NHC catalysed Carbonylative Suzuki cross-coupling reactions of aryl halides and arylboronic acids and its application towards the synthesis of biologically active 4-quinolone scaffolds. *Appl. Organomet. Chem.*, 32: e4173.

[19] Venkataraman, S., Barange, D. K. and Pal, M. (2006). One-pot synthesis of 2-substituted furo[3,2-c]quinolines via tandem coupling–cyclization under Pd/C-copper catalysis. *Tetrahedron Lett.* 47: 7317-7322.

[20] Mugnaini, C., Falciani, C., Rosa, M. De., Brizzi, A., Pasquini, S. and Corelli, F. (2011). *Tetrahedron*, 67, 5776-5783.

[21] Pinto, J., Silva, V. L. M., Santos, L. M. N. B. F. and Silva, A. M. S. (2016). Synthesis of (E)-3-Styrylquinolin-4(1*H*)-ones in Water by Ohmic Heating: a Comparison with Other Methodologies *Eur. J. Org. Chem.*, 2888–2896.

[22] Audisio, D., Messaoudi, S., Peyrat, J.-F., Brion, J.-D. and Alami, M. (2011). A General Copper Powder-Catalyzed Ullmann-Type Reaction of 3-Halo-4(1*H*)-quinolones With Various Nitrogen-Containing Nucleophiles *J. Org. Chem.*, 76: 4995-5005.

[23] Ghosh, P. and Das, S. (2018). Ligand Free Approach for the Copper(II)-Mediated C-NH_2 Arylation of 4-Quinolone Derivatives Under Ambient Condition. *Chemistry Select*, 3: 8624-8627.

[24] Messaoudi, S., Brion, J.-D. and Alami, M. (2012). Palladium-Catalyzed Decarboxylative Coupling of Quinolinone-3-Carboxylic Acids and Related Heterocyclic Carboxylic Acids with (Hetero)aryl Halides. *Org. Lett.*, 14: 1496-1499.

[25] Reddy, K. H. V., Brion, J.-D., Messaoudi, S., and Alami, M. (2016). Synthesis of Biheterocycles based on Quinolinone, Chromone and Coumarin Scaffolds by Palladium-Catalyzed Decarboxylative Couplings. *J. Org. Chem.*, 81: 424–432.

[26] Azad, C. S., Balaramnavar, V. M., Khan, I. A., Doharey, P. K., Saxena, J. K. and Saxena, A. K. (2015). Operative conversions of 3-carboxy-4-quinolones into 3-nitro-4- quinolones via ipso-nitration: Potential antifilarial agents as inhibitor of Brugia malayi thymidylate kinase. *RSC Adv.*, *5*: 82208-82214.

[27] Azad, C. S. and Narula, A. K. (2016). An operational transformation of 3-carboxy-4-quinolones into 3- nitro-4-quinolones via ipso-nitration using polysaccharide supported copper nanoparticles: synthesis of 3-tetrazolyl bioisosteres of 3-carboxy-4-quinolones as antibacterial agents. *RSC Adv.*, 6: 19052-19059.

[28] Xia, C., Wei, Z., Yang, Y., Yu, W., Liao, H., Shen, C. and Zhang, P. (2016). Palladium-Catalyzed Thioetherification of Quinolone Derivatives via Decarboxylative C-S Cross-Couplings. *Chem. Asian. J.*, 11: 360-366.

[29] Li, M., Li L. and Ge, H. (2010). Direct C-3-Alkenylation of Quinolones via Palladium-Catalyzed CH Functionalization67. *Adv. Synth. Catal.*, 352: 2445-2449.

[30] Kwon, S., Kang, D. and Hong, S. (2015). Rh(I)-Catalyzed Site-Selective Decarbonylative Alkenylation and Arylation of Quinolones under Chelation Assistance. *Eur. J. Org. Chem.*, 3671-3678.

[31] Banerjee, A., Santra, S. K., Mohanta, P. R. and Patel, B. K. (2015). Ruthenium(II) Catalyzed Regiospecific C−H/O−H Annulations of Directing Arenes via Weak Coordination. *Org. Lett.,* 17: 5678-5681.

[32] Kang, D. and Hong S., (2015). Rh(III) and Ru(II)-Catalyzed Site-Selective C−H Alkynylation of Quinolones. *Org. Lett.*, 17: 1938-1941.

[33] Das, D. and Samanta, R. (2018). Iridium Catalyzed Regiocontrolled Direct Amidation of Isoquinolones and Pyridones. *Adv. Synth. Catal*., 360: 379-384.

[34] Shin, S., Kim, Y., Kim, K. and Hong, S. (2014). A copper-mediated cross-coupling approach for the synthesis of 3-heteroaryl quinolone and related analogues. *Org. Biomol. Chem.*, 12: 5719-5726.

[35] Mehra, M. K., Sharma, S., Rangan, K. and Kumar, D. Substrate or Solvent-Controlled PdII-Catalyzed Regioselective Arylation of Quinolin-4(1*H*)-ones. (2020). Using Diaryliodonium Salts: Facile Access to Benzoxocine and Aaptamine Analogues. *Eur. J. Org. Chem*., 2409-2413.

[36] Han, X., Yue, Z., Zhang, X., He, Q. and Yang, C. (2013). Copper-Mediated, Palladium-Catalyzed Cross-Coupling of 3-Iodochromones, Thiochromones, and Quinolones with Ethyl Bromodifluoroacetate. *J. Org. Chem.,* 78: 4850-4856.

[37] An, Z.-Y., Yan, Y. -Y., Peng, D., Ou, T. -M., Tan, J. -H., Huang, S. L., An, L. -K., Gu, L. -Q. and Huang, Z. -S. (2010). *Eur. J. Med. Chem.,* 45: 3895-3903.

[38] Li, X., Bian, Y., Chen, X., Zhang, H., Wang, W., Ren, S., Yang, X., Lu, C., Chen, C. and Peng, J. (2019). Tunable synthesis of quinolinone-fused isoquinolines through sequential one-pot nucleophilic addition and palladium-catalyzed intramolecular C–H alkenylation. *Org. Biomol. Chem.*, 17: 321-332.

[39] Arasakumar, T., Shyamisivappan, S., Gopalan, S., Ata, A. and Mohan, P.S. (2019). One-Pot Approach to Pyrido-4-phenanthridinones by Palladium Catalyzed Annulation of 4-Quinolones with 2-Bromobenzyl Bromides. *Synlett.*, 30: 63-68.

[40] Shintani, R., Yamagami, T., Kimura, T. and Hayashi, T. (2005). A New Entry of Nucleophiles in Rhodium-Catalyzed Asymmetric 1,4-Addition Reactions: Addition of Organozinc Reagents for the Synthesis of 2-Aryl-4-piperidones.*Org. Lett.,* 7: 5317-5319.

[41] Guo, Y. and Harutyunyan, S. R. (2019). Highly enantioselective catalytic addition of Grignard reagents to N-heterocyclic acceptors. *Angew. Chem. Int. Ed*., 58: 12950-12954.

[42] He, B., Phansavath, P., Ratovelomanana-Vidal, V. (2020). Rhodium-catalyzed asymmetric transfer hydrogenation of 4- quinolone derivatives. *Org. Chem. Front*., 7: 975-979.

[43] Ji, F., Li, X., Wu, W. and Jiang, H. (2014). Palladium-Catalyzed Oxidative Carbonylation for the Synthesis of Polycyclic Aromatic Hydrocarbons (PAHs). *J. Org. Chem*., 79: 11246-1253.

Chapter 3

Screened Exchange Interactions at the Interface of Transition Metals

Adam B. Cahaya*

Department of Physics, Department of Mathematics and Natural Sciences, Universitas Indonesia, Depok, Indonesia

Abstract

Exchange interaction at the bulk of transition metal is well described with Hund's J parameter. For bare Coulomb interactions, J can be analytically obtained. However, the value is too large. Although J is often treated as independent to the orbital-pairs, Racah parameters suggest that the value depends on the orbital pairs. In this chapter, we modify the Racah parameters using a Yukawa-type screened Coulomb interaction and show that in a strong screening limit, J is independent of the orbital pair.

On the other hand, the exchange parameter at the interface of transition metal is often described with s-d exchange interaction. While interfacial interaction seems to be limited only by the dirtiness of the interface, we show that in strong screening limit, the exchange interaction at the interface is limited by the electron-electron repulsive interaction.

Keywords: exchange parameter, screened Coulomb interaction, Yukawa potential

* Corresponding Author's Email: adam@sci.ui.ac.id.

In: Transition Metals - An Overview
Editor: Veronica J. Avila
ISBN: 979-8-88697-646-5

© 2023 Nova Science Publishers, Inc.

1. Introduction

The wave function $\psi_{3d,m}(\mathbf{r}) = R_{3d}(r)Y_2^m(\hat{\mathbf{r}})$ of d-electron that characterizes transition metals is the solution of Schrödinger equation for a single electron. Here $Y_{lm}(\theta, \varphi)$ is the spherical harmonics, $R_{3d}(r)$ is the radial wave function

$$R_{3d}(r) = \sqrt{\frac{8\zeta^7}{45}}r^2 e^{-\zeta r},$$

$$\zeta = \frac{(Z - s)}{3a_0},$$

that depends on the shielded nuclei charge $Z - s$ (Slater 1930). a_0 is the Bohr radius. The shielding arises from the screening of nuclei by other electrons (Pauling and Sommerfeld 1927). Since electrons are fermions, the solution of Schrödinger for two electrons systems takes the form of Slater determinant (Slater 1929)

$$\psi(\mathbf{r}_1, \mathbf{r}_2) = \frac{1}{\sqrt{2}}\begin{vmatrix} \psi_1(\mathbf{r}_1) & \psi_1(\mathbf{r}_2) \\ \psi_2(\mathbf{r}_1) & \psi_2(\mathbf{r}_2) \end{vmatrix}$$

Consequently, the electrostatic interaction between two electrons

$$\langle V \rangle = \int d^3r_1 d^3r_2\, \psi^*(\mathbf{r}_1, \mathbf{r}_2)V(|\mathbf{r}_1 - \mathbf{r}_2|)\psi(\mathbf{r}_1, \mathbf{r}_2) = U - J$$

has direct

$$U = \int d^3r_1 d^3r_2\, \psi_1^*(\mathbf{r}_1)\psi_2^*(\mathbf{r}_2)V(|\mathbf{r}_1 - \mathbf{r}_2|)\psi_1(\mathbf{r}_1)\psi_2(\mathbf{r}_2),$$

and exchange terms

$$J = \int d^3r_1 d^3r_2\, \psi_1^*(\mathbf{r}_2)\psi_2^*(\mathbf{r}_1)V(|\mathbf{r}_1 - \mathbf{r}_2|)\psi_1(\mathbf{r}_1)\psi_2(\mathbf{r}_2).$$

The exchange interaction has been shown to be responsible for ferromagnetism (Heisenberg 1928).

The exchange energy between electrons in 3d orbitals is pair-dependent

$$J_{ab} = \int d^3r_1 d^3r_2\, R_{3d}^2(r_1)R_{3d}^2(r_2)V(|\mathbf{r}_1 - \mathbf{r}_2|)Y_2^{a*}(\hat{\mathbf{r}}_2)Y_2^{b*}(\hat{\mathbf{r}}_1)Y_2^{a}(\hat{\mathbf{r}}_1)Y_2^{b}(\hat{\mathbf{r}}_2)$$

J for intraorbital exchange of two electrons with parallel spins is responsible for Hund's coupling (Georges, Medici, and Mravlje 2013). J_{mn} separated to radial and angular integrations using spherical harmonic expansion of bare Coulomb potential

$$\frac{1}{4\pi|\mathbf{r}_1 - \mathbf{r}_2|} = \sum_{l=0}^{\infty} \frac{1}{2l+1}\frac{r_<^l}{r_>^{l+1}} \sum_{m=-l}^{l} Y_l^m(\hat{\mathbf{r}}_1)\, Y_l^{m*}(\hat{\mathbf{r}}_2),$$

where $r_< = \min(r_1, r_2)$ and $r_> = \max(r_1, r_2)$. Such that

$$J_{ab} = \frac{e^2}{\varepsilon_0}\sum_{lm}\frac{F^l}{2l+1}\int d\Omega_1 \int d\Omega_2\, Y_l^m(\hat{\boldsymbol{r}}_1)\, Y_l^{m*}(\hat{\boldsymbol{r}}_2)Y_2^{a*}(\hat{\boldsymbol{r}}_2)Y_2^{b*}(\hat{\boldsymbol{r}}_1)Y_2^{a}(\hat{\boldsymbol{r}}_1)Y_2^{b}(\hat{\boldsymbol{r}}_2),$$

can be written in term of Slater integrals

$$F^l = \int_0^{\infty} r_1^2 dr_1 \int_0^{\infty} r_2^2 dr_2\, R_{3d}^2(r_1)R_{3d}^2(r_2)\frac{r_<^l}{r_>^{l+1}}.$$

The expression of J_{mn} can be further simplified if Y_2^m
fact, the 3d angular wave function of transition metal compounds are usually real-valued due to crystal field (van Vleck 1932).

A transition metal ion in a compound is subjected electrostatic fields from ions in its crystal environment. As a result, 3d electrons no longer have rotational symmetry and Y_l^m
equation, but are replaced by their symmetry-adapted linear combinations. For crystals with cubic symmetries, magnetic ions tend to occupy octahedral and tetrahedral sites, the symmetry-adapted angular wave functions are triply generated t_{2g} orbitals

$$Y_{xz} = \frac{1}{\sqrt{2}}(Y_2^{-1} - Y_2^1) = \sqrt{\frac{15}{4\pi}}\frac{xz}{r^2}$$

$$Y_{yz} = \frac{i}{\sqrt{2}}(Y_2^{-1} + Y_2^1) = \sqrt{\frac{15}{4\pi}}\frac{yz}{r^2}$$

$$Y_{xy} = \frac{i}{\sqrt{2}}(Y_2^{-2} - Y_2^2) = \sqrt{\frac{15}{4\pi}}\frac{xy}{r^2}$$

and doubly generated e_g

$$Y_{z^2} = Y_2^0 = \sqrt{\frac{5}{16\pi}}\frac{3z^2 - r^2}{r^2}$$

$$Y_{x^2-y^2} = \frac{1}{\sqrt{2}}(Y_2^{-2} + Y_2^2) = \sqrt{\frac{5}{16\pi}}\frac{3z^2 - r^2}{r^2}$$

Substituting the real-valued angular wave functions, the exchange interaction between electrons with the real-valued orbitals are linear combinations of J and ΔJ

$$J = \frac{1}{2}\left(J_{e_g e_g} + J_{t_{2g} t_{2g}}\right) = \frac{e^2}{14\varepsilon_0}\left(F^2 + \frac{5}{9}F^4\right) = \frac{611 e^2 \zeta}{46080\varepsilon_0} \approx \frac{e^2}{22\varepsilon_0 \langle r \rangle_{3d}},$$

$$\Delta J = \frac{1}{2}\left(J_{e_g e_g} - J_{t_{2g} t_{2g}}\right) = \frac{e^2}{98\pi\varepsilon_0}\left(F^2 - \frac{5}{9}F^4\right) = \frac{143 e^2 \zeta}{161280\varepsilon_0} \approx \frac{e^2}{322\varepsilon_0 \langle r \rangle_{3d}},$$

as summarized in Table 1 (Eder 2012; Oles and Stollhoff 1984). Here we use the related Slater integrals

$$F^2 = \frac{2093}{15360}\zeta \simeq \frac{0.477}{\langle r \rangle_{3d}},$$

$$F^4 = \frac{91}{1024}\zeta \simeq \frac{0.311}{\langle r \rangle_{3d}}$$

$\langle r \rangle_{3d} = 7/(2\zeta)$ is the averaged radius of 3d electron. One can see that $\Delta J \approx J/15$ is an order of magnitude smaller than J. As a result, ΔJ J_{ab} is often neglected. Assuming that the average radius is in the order of Bohr radius $\langle r \rangle_{3d} \sim a_0$, the exchange parameter

$$J_{ab} \sim \begin{cases} 16 \text{ eV} & a \neq b, \\ 0, & a = b, \end{cases}$$

is larger than the actual value $J_{a \neq b} \sim 0.4$ eV for V, Cr, Mn (Vaugier, Jiang, and Biermann 2012). For a realistic value, one needs to consider a screened Coulomb interaction.

Screened Coulomb interaction is also required to accurately describe the properties of transition metal-based materials, such as metal - insulator transition (Imada, Fujimori, and Tokura 1998), superconductivity (Barnes 2004), and ferroelectricity (Wang et al. 2012). The mechanism of electric field screening is discussed in Section 1. In section 2, we discuss the exchange interaction that arises from screened Coulomb potentials in the bulk of a transition metal. Section 3 focuses on the exchange interaction at magnetic interfaces. The interfacial exchange interaction shows that screened Coulomb potential is measurable in an interfacial phenomenon.

Table 1. Intraorbital exchange $\boldsymbol{J_{ab}}$ for real-valued 3d orbitals. $\boldsymbol{J_{e_g e_g} = J - \Delta J}$ and $\boldsymbol{J_{t_{2g} t_{2g}} = J + \Delta J}$

$a \backslash b$	xy	yz	xz	$x^2 - y^2$	$3z^2 - r^2$
xz	0	$J_{t_{2g}t_{2g}}$	$J_{t_{2g}t_{2g}}$	$J - \Delta J$	$J - 5\Delta J$
yz	$J_{t_{2g}t_{2g}}$	0	$J_{t_{2g}t_{2g}}$	$J - \Delta J$	$J - 5\Delta J$
xy	$J_{t_{2g}t_{2g}}$	$J_{t_{2g}t_{2g}}$	0	$J - 7\Delta J$	$J + \Delta J$
$x^2 - y^2$	$J - \Delta J$	$J - \Delta J$	$J - 7\Delta J$	0	$J_{e_g e_g}$
z^2	$J - 5\Delta J$	$J - 5\Delta J$	$J + \Delta J$	$J_{e_g e_g}$	0

2. Theories of Electric Field Screening

2.1. Thomas-Fermi Theory for a Many-Electrons System

The screening of Coulomb interaction was first studied by Thomas and Fermi for describing many-electron problems (Lieb and Simon 1977). The screened Coulomb interaction $\bar{V}(\mathbf{r})$ depends on the electron density via Poisson equation

$$\nabla^2 \bar{V}(\mathbf{r}) = \frac{e^2}{\varepsilon_0} n(\mathbf{r}).$$

Our $\bar{V}$ has the same dimension as energy. Thomas-Fermi theory assumes that the optimum electron density gives the minimum total energy

$$E = \int d^3r \, \epsilon_{kinetic} + \frac{1}{2}\int d^3r_1 d^3r_2 \, n(\mathbf{r}_1) n(\mathbf{r}_2) V_e(|\mathbf{r}_1 - \mathbf{r}_2|).$$

The second term is the electron - electron interaction. Factor 1/2 is to avoid double counting. The first term is the kinetic energy density of a free electronic system

$$\epsilon_{kinetic} = 2\int_0^{k_F} \frac{d^3k}{(2\pi)^3} \frac{\hbar^2 k^2}{2m} = \frac{\hbar^2 k_F^5}{10\pi^2 m}.$$

Factor 2 in front of the integral is the spin degeneracy. The Fermi wavevector k_F is also related to electron density, which is assumed to be the number of

$$n = 2\int_0^{k_F} \frac{d^3k}{(2\pi)^3} = \frac{k_F^3}{3\pi^2}.$$

Substituting $k_F[n]$ to $\epsilon_{kinetic}$, we arrive at the expression of $E[n(\mathbf{r})]$ as a functional of electron density $n(\mathbf{r})$ (March 1983)

$$E[n(\mathbf{r})] = C\int d^3r \, [n(\mathbf{r})]^{\frac{5}{3}} + \frac{1}{2}\int d^3r_1 d^3r_2 \, n(\mathbf{r}_1) n(\mathbf{r}_2) V_e(|\mathbf{r}_1 - \mathbf{r}_2|),$$

$C = \hbar^2(3\pi)^{\frac{5}{3}}/(10\pi m)$. By imposing total charge conservation

$$\int d^3r\, n(\mathbf{r}) = N,$$

$n(r)$ can be determined by minimizing

$$E - \mu N$$

Using calculus of variation, the optimum $n(\mathbf{r})$ can be shown to fulfill

$$0 = \frac{5}{3}C[n(\mathbf{r})]^{\frac{2}{3}} + \int d^3r'\, n(\mathbf{r}')V_e(|\mathbf{r} - \mathbf{r}'|) - \mu.$$

The third term is screened potential $\bar{V}(\mathbf{r})$. $\bar{V}(\mathbf{r})$ can be found by substituting

$$n(\mathbf{r}) = \left(\frac{3}{5C}\right)^{\frac{3}{2}} [\mu - \bar{V}(\mathbf{r})]^{\frac{3}{2}},$$

back to Poisson equation

$$\nabla^2\bar{V}(\mathbf{r}) = \frac{e^2}{\varepsilon_0}\left(\frac{3}{5C}\right)^{\frac{3}{2}} [\mu - \bar{V}(\mathbf{r})]^{\frac{3}{2}}$$

By setting $\mu = 0$, one can find the Thomas-Fermi equation for screening factor $\phi \propto r\bar{V}(\mathbf{r})$ of a single atom (Plindov and Pogrebnya 1987; Fernández 2008)

$$\frac{\partial^2\phi}{\partial x^2} = \frac{\phi^{\frac{3}{2}}}{\sqrt{x}}.$$

While Thomas-Fermi theory fails to the explain stability of negative ions and molecular binding, it is the oldest density functional theory (DFT) (Solovej 2016; Morgan III 2006). This functional treatment of electron density in the Thomas-Fermi theory is the precursor of the modern DFT (R. O. Jones and Gunnarsson 1989). In the Kohn-Sham density functional theory, Thomas-

Fermi theory is improved by using the quantum kinetic energy operator (King and Handy 2000)

$$E_{kinetic} = \int d^3r\, \psi^*(\mathbf{r}) \left(-\frac{\hbar^2}{2m} \nabla^2 \right) \psi(\mathbf{r}).$$

The total energy $E(\rho(\mathbf{r}))$ of Kohn-Sham equation also includes the correction from the exchange correlation correction. In recent decades, the refinements of DFT lead to its success on predicting energy levels, charge density and subsequent properties of electrons in realistic materials (Burke 2012). The significance of DFT is reflected in the 1998 Nobel Prize in Chemistry, which was awarded to Walter Kohn for the development of DFT (Kohn 1999) and John A. Pople for the development of computational methods in quantum chemistry (Pople 1999).

Our system of interest is a transition metal system with many electrons $N \to \infty$. In this case, $\bar{V}(\mathbf{r})$ can be assumed to be much smaller than the chemical potential μ, such that

$$\nabla^2 \bar{V}(\mathbf{r}) \approx \frac{e^2}{\varepsilon_0} \left[n_0(\mathbf{r}) - \bar{V}(\mathbf{r}) \frac{\partial n}{\partial \mu} \right]$$

here $\rho_0(\mathbf{r})$ is the electron density without screening. Defining

$$\lambda = \sqrt{\frac{e^2}{\varepsilon_0} \frac{\partial n}{\partial \mu}}$$

as the Thomas-Fermi screening wavevector, one can see that the solution of

$$\nabla^2 \bar{V}(\mathbf{r}) + \lambda^2 \bar{V}(\mathbf{r}) = \frac{e^2}{\varepsilon_0} n_0(\mathbf{r})$$

for $n_0(\mathbf{r}) = \delta(\mathbf{r})$ take the form of Yukawa potential

$$\bar{V}(r) = \frac{e^2}{4\pi\varepsilon_0 r} e^{-\lambda r}.$$

Taking a look at the screened Coulomb interaction in momentum space

$$\bar{V}(q) = \frac{e^2/\varepsilon_0}{q^2+\lambda^2} \equiv \frac{V(q)}{\varepsilon_r(q)},$$

one can see that the correction due to screening of electrostatic interaction can be characterized by momentum-dependent relative permittivity (Resta 1977)

$$\varepsilon_r(q) = 1 + \frac{\lambda^2}{q^2},$$

For large λ, $\lim_{\lambda \gg q} \bar{V}(q) = e^2/\varepsilon_0\lambda^2$ become independent to q and $\bar{V}(r)$ becomes a delta function.

2.2. Lindhard Dielectric Function

The screening of electric fields can be understood as additional charge density $\Delta n(\mathbf{r})$

$$\nabla^2 \bar{V}(\mathbf{r}) = \frac{e^2}{\varepsilon_0}\big(\delta(\mathbf{r}) + \Delta n(\mathbf{r})\big)$$

$\delta n(\mathbf{r})$ can be found using linear response theory (Kim 1999a)

$$\Delta n(\mathbf{r}) = -\int d^3r\, \chi_c(\mathbf{r}-\mathbf{r}', t-t')\bar{V}(\mathbf{r}).$$

here χ_c is the charge susceptibility. Its Fourier transform can be evaluated using random phase approximation

$$\chi_c(\mathbf{q},\omega) = \lim_{\eta\to 0}\sum_{\mathbf{k}} \frac{n_{\mathbf{k}} - n_{\mathbf{k+q}}}{E_{\mathbf{k+q}} - E_{\mathbf{k}} + \hbar\omega + i\eta}.$$

$\eta \to 0$ ensures that χ_c is a retarded response. Here $n_{\mathbf{k}}$ is the Fermi-Dirac distribution of electron with wave vector **k** and energy $E_{\mathbf{k}}$
momentum-dependent relative permittivity from the Fourier transform of the Poisson equation

$$-q^2\bar{V}(\mathbf{q}) = \frac{e^2}{\varepsilon_0}\left(1 + \chi_c(\mathbf{q})\bar{V}(\mathbf{q})\right)$$

$$\bar{V}(\mathbf{q}) = \frac{e^2}{\varepsilon_0\left(q^2 + \frac{e^2}{\varepsilon_0}\chi_c(\mathbf{q},\omega)\right)} \equiv \frac{V(\mathbf{q})}{1 + \varepsilon_r(\mathbf{q},\omega)}$$

The relative permittivity is also known as Lindhard dielectric function (Lindhard 1954; Andrade-Neto 2016)

$$\varepsilon_r(\mathbf{q},\omega) = 1 + \frac{e^2}{\varepsilon_0 q^2}\chi_c(\mathbf{q},\omega) = 1 + V(\mathbf{q})\lim_{\eta\to 0}\sum_{\mathbf{k}}\frac{n_{\mathbf{k}} - n_{\mathbf{k+q}}}{E_{\mathbf{k+q}} - E_{\mathbf{k}} + \hbar\omega + i\eta}$$

Since $\varepsilon_r = 1 + \lambda^2/q^2$, one can see that the charge susceptibility is related to the screening constant λ

$$\lambda^2 = \lim_{q\to 0}\frac{e^2}{\varepsilon_0}\chi_c(\mathbf{q},0).$$

The static and long wavelength limit of the related charge susceptibility of χ_e can be evaluated to be

$$\chi_e(\mathbf{0},0) = \sum_{\mathbf{k}}\frac{n_{\mathbf{k}} - n_{\mathbf{k+q}}}{E_{\mathbf{k+q}} - E_{\mathbf{k}}} = -\sum_{\mathbf{k}}\frac{\partial n_{\mathbf{k}}}{\partial E_{\mathbf{k}}} = N_F.$$

Such that $\lambda = \sqrt{e^2 N_F/\varepsilon_0}$.

For an interacting electron system, such as the conduction electrons of heavy transition metals, χ_c can be studied using interacting electron Hamiltonian (Doniach and Sondheimer 1998a).

$$H = \sum_{\mathbf{k}} E_{\mathbf{k}}\, n_{\mathbf{k}} + \sum_{\mathbf{k}} U n_{\mathbf{k},\uparrow} n_{-\mathbf{k},\downarrow}.$$

here, U is Hubbard parameter that characterizes the direct Coulomb interaction between conduction electrons. Using random phase approximation, $\chi_c(\mathbf{q},\omega)$ has been shown to be reduced by positive U (Kim 1999a)

$$\chi_c = \frac{\chi_c^0}{1 + UN_F}.$$

See Appendix for derivation. Therefore, $\lambda^2 \propto 1/(1 + UN_F)$ of interacting electrons is smaller than the non-interacting system.

2.3. Friedel Theory for an Interacting Electron System

In the strong screening limit,

$$\lim_{\lambda \gg q} \vec{V}(\mathbf{q}) = \frac{e^2}{\varepsilon_0 \lambda^2}$$

is independent of q and $\vec{V}(\mathbf{q})$ becomes a delta function. The charge distribution Δn is then directly determined by χ_c

$$\Delta n(\mathrm{r}) = -\frac{e^2}{\varepsilon_0 \lambda^2} \chi_c(\mathrm{r}, 0) = -\frac{1}{N_F} \sum_{\mathrm{kq}} e^{i\mathrm{q}\cdot\mathrm{r}} \frac{n_{\mathrm{k}} - n_{\mathrm{k+q}}}{E_{\mathrm{k+q}} - E_{\mathrm{k}}}$$

By taking the integral of $\Delta n(\mathbf{r})$ over the whole space, we can see that the charge is totally screened

$$\int d^3 r \, \Delta n(\mathbf{r}) = -\frac{1}{N_F} \sum_{\mathbf{kq}} \frac{n_{\mathbf{k}} - n_{\mathbf{k+q}}}{E_{\mathbf{k+q}} - E_{\mathbf{k}}} \int d^3 r \, e^{i\mathbf{q}\cdot\mathbf{r}} = -1.$$

It is interesting to see, the spatial dependency of Δn. The summation can be evaluated by substituting $\mathbf{k} + \mathbf{q} = \mathbf{p}$ and replacing the summation by integral

$$\begin{aligned}\Delta n(\mathbf{r}) &= -\frac{1}{N_F} \int d^3 k d^3 p \, e^{i\mathbf{p}\cdot\mathbf{r}} e^{-i\mathbf{k}\cdot\mathbf{r}} \frac{n_{\mathbf{k}} - n_{\mathbf{p}}}{E_{\mathbf{p}} - E_{\mathbf{k}}} \\ &= -\frac{1}{N_F} \int d^3 k d^3 p \, e^{i\mathbf{p}\cdot\mathbf{r}} e^{-i\mathbf{k}\cdot\mathbf{r}} \frac{2 n_{\mathbf{k}}}{E_{\mathbf{p}} - E_{\mathbf{k}}}\end{aligned}$$

Since the system is isotropic

$$\Delta n(r) = -\frac{2m}{\hbar^2 N_F}\frac{(4\pi)^2}{(2\pi)^6}\int_0^{k_F} k^2 dk \frac{\sin kr}{kr}\int_0^{\infty} p^2 dp \frac{\sin pr}{pr}\frac{1}{p^2 - k^2}$$
$$= -\frac{1}{8\pi r^3}\left(\frac{\sin 2k_F r}{2k_F r} - \cos 2k_F r\right).$$

The spatial oscillation of $\Delta n(r)$ is called Friedel oscillation (Bena 2016).

3. Screened Exchange Interactions in the Bulk of Transition Metal

3.1. Screened Hund Exchange

To describe J using the screened potential, we can use the following spherical harmonics expansion (Cahaya, Azhar, and Majidi 2021)

$$\frac{e^{-\lambda|\mathbf{r}_1 - \mathbf{r}_2|}}{4\pi|\mathbf{r}_1 - \mathbf{r}_2|} = \sum_{l=0}^{\infty} \lambda\, i_l(\lambda r_<) k_l(\lambda r_>) \sum_{m=-l}^{l} Y_l^m(\hat{\mathbf{r}}_1)\, Y_l^{m*}(\hat{\mathbf{r}}_2).$$

here $i_n(x)$ and $k_n(x)$ are modified spherical Bessel functions of the first and second kind, respectively. Due to the spherical harmonic expansion, one can see that the Slater integration can be modified by replacing $r_<^l / r_>^{l+1}$ with $(2l+1)\lambda i_l(\lambda r_<) k_l(\lambda r_>)$. Therefore

$$\bar{J} = \frac{5\lambda e^2}{14\varepsilon_0}\int_0^{\infty} r_1^2 dr_1 \int_0^{\infty} r_2^2 dr_2\, R_{3d}^2(r_1) R_{3d}^2(r_2)[i_2(\lambda r_<)k_2(\lambda r_>) + \lambda i_4(\lambda r_<)k_4(\lambda r_>)]$$

$$\Delta\bar{J} = \frac{5\lambda e^2}{98\varepsilon_0}\int_0^{\infty} r_1^2 dr_1 \int_0^{\infty} r_2^2 dr_2\, R_{3d}^2(r_1) R_{3d}^2(r_2)[i_2(\lambda r_<)k_2(\lambda r_>) - \lambda i_4(\lambda r_<)k_4(\lambda r_>)].$$

Numerical evaluations indicate that $\bar{J}$ and $\Delta\bar{J}$ are monotonically decreasing functions of λ/ζ. For strong screening limits the values converge (Cahaya, Azhar, and Majidi 2021)

$$\bar{J} \simeq \frac{7e^2}{256\varepsilon_0}\frac{\zeta^3}{\lambda^2}$$

$$\Delta\bar{J} \simeq \frac{5e^2}{288\varepsilon_0}\frac{\zeta^5}{\lambda^4}$$

One can see that in a strong screening limit $\lim_{\lambda \gg \zeta} \Delta\bar{J} \ll \bar{J}$. The strong screening limit is useful for discussing exchange interaction between localized and itinerant electrons.

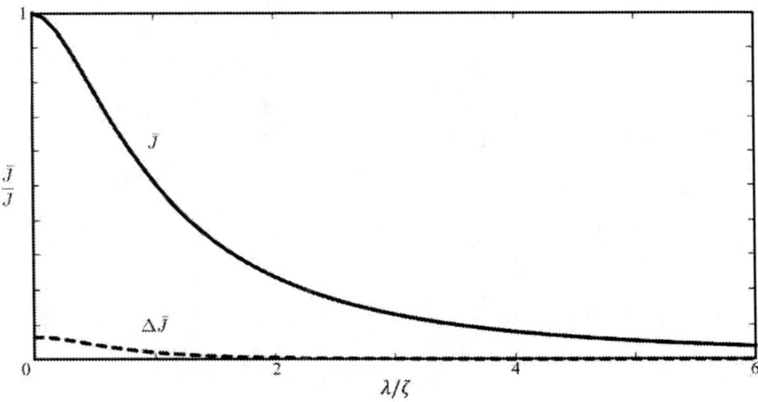

Figure 1. $\bar{J}$ (solid line) and $\Delta\bar{J}$ (dashed line) are monotonically decreasing functions of λ/ζ. The values are normalized to $J_0 = e^2/(22\varepsilon_0\langle r\rangle_{3d}) \sim 16\, eV$.

3.2. Screened $s - d$ Exchange Interaction

The exchange interaction can also happen between electrons of different orbitals. Let us focus on exchange interaction between s and d electrons. Since s-electron is much less localized than d electron, its wave function can be approximated using plane wave

$$J_{sd} = \frac{1}{V_0}\int d^3r_1 d^3r_2\, e^{i(\mathbf{k}_1\cdot\mathbf{r}_1 - \mathbf{k}_2\cdot\mathbf{r}_2)}\psi_{3d}^*(\mathbf{r}_1)\psi_{3d}(\mathbf{r}_2)V(|\mathbf{r}_1 - \mathbf{r}_2|),$$

where $|\mathbf{k}_1| = |\mathbf{k}_2| \approx k_F$ (Kondo 1962). k_F is the Fermi wavevector of the conduction electron. V_0
unit volume. For bare Coulomb interaction the leading order of the exchange interaction is

$$J_{sd} = \frac{e^2}{\varepsilon_0 V_0}\int dr_1 dr_2\, r_1^2 r_2^2 j_0(k_F r_1) j_0(k_F r_2) R_{3d}(r_1) R_{3d}(R_2)\frac{r_<^2}{5r_>^3}$$

$$= \frac{e^2}{4\pi\varepsilon_0 V_0}\frac{27 - 20\left(\frac{k_F}{\zeta}\right)^2 + 27\left(\frac{k_F}{\zeta}\right)^4 + 15\left(\frac{k_F}{\zeta}\right)^6 + 6\left(\frac{k_F}{\zeta}\right)^8 + \left(\frac{k_F}{\zeta}\right)^{10}}{15\left(1 + \left(\frac{k_F}{\zeta}\right)^2\right)^7}.$$

For localized limit of d-electron $\lim_{\zeta \gg k_F} J_{sd} = 9e^2(5\varepsilon_0 V_0 \zeta^2)^{-1} = 36e^2\langle r\rangle_{3d}^2(245\varepsilon_0 V_0)^{-1}$. The screened exchange constant can be calculated by replacing the bare Coulomb potential with $\bar{V}(r)$

$$\bar{J}_{sd} = \frac{e^2\lambda}{\varepsilon_0 V_0}\int r_1^2 r_2^2 dr_1 dr_2\, j_0(k_F r_1) j_0(k_F r_2) R_{3d}(r_1) R_{3d}(r_2) i_2(\lambda r_<) k_2(\lambda r_>).$$

The localized limit of d-electron (Cahaya and Majidi 2021) is

$$\lim_{\zeta \gg k_F} \bar{J}_{sd} = \frac{e^2}{\varepsilon_0 V_0}\frac{1 + 8\frac{2\zeta}{\lambda} + 27\left(\frac{2\zeta}{\lambda}\right)^2 + 48\left(\frac{2\zeta}{\lambda}\right)^3 + \frac{219}{5}\left(\frac{2\zeta}{\lambda}\right)^4 + \frac{72}{5}\left(\frac{2\zeta}{\lambda}\right)^5 + \frac{9}{5}\left(\frac{2\zeta}{\lambda}\right)^6}{\lambda^2\left(1 + \frac{2\zeta}{\lambda}\right)^8}.$$

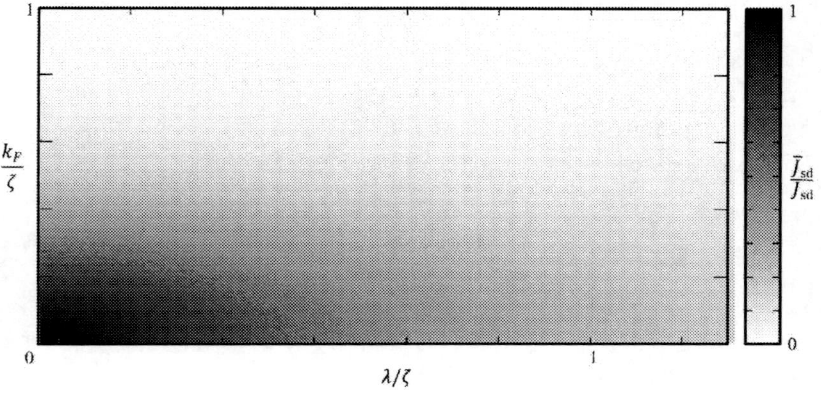

Figure 2. s-d exchange interaction constant $\bar{J}_{sd}$ as a function of λ/ζ and k_F/ζ. The values is normalized to the value for $k_F = 0 = \lambda$.

4. Exchange in Magnetic Interface of Transition Metals

In this section we focus on the exchange interaction at the interface between ferromagnet and nonmagnetic metal. At the magnetic interface, there are localized spins from d-electrons of ferromagnet and conduction electrons from s-electrons (Cahaya 2018). For discussion in this section, we will focus on the strong screening limit of $\bar{J}_{sd}$

$$\lim_{\lambda \gg \zeta \gg k_F} \bar{J}_{sd} = \frac{e^2}{V_0 \lambda^2}$$

When the interface is metallic, the dielectric function can also be described by the Lindhard theory (Miraglia and Gravielle 2002; Isihara 1998). When the nonmagnetic metal is a heavy metal, we also need to treat the s-electron as an interacting electron system

4.1. Interlayer RKKY Interaction

In the strong screening limit, the s-d exchange interaction Hamiltonian can be written

$$\begin{aligned}\lim_{\lambda \gg \zeta} H_{s-d} &= -\lim_{\lambda \gg \zeta} \sum_{\mathbf{k}_1 \mathbf{k}_2} \bar{J}_{sd}\, \mathbf{S}_{\mathrm{d}} \cdot \mathbf{s}(\mathbf{k}_2 - \mathbf{k}_1) \\ &= -\frac{e^2}{\varepsilon_0 V_0} \int d^3r\, |\psi_{3d}(\mathbf{r})|^2 \mathbf{S}_{\mathrm{d}} \cdot \mathbf{s}(\mathbf{r})\end{aligned}$$

as an exchange magnetic field acting on the magnetization of conduction spin $(e/m_e)\mathbf{s}$

$$\mathbf{B}_{sd}(\mathbf{r}) = \frac{em_e}{\varepsilon_0 V_0} |\psi_{3d}(\mathbf{r})|^2 \mathbf{S}_{\mathrm{d}} \simeq \mathbf{S}_{\mathrm{d}} \frac{em_e}{\varepsilon_0 V_0} \delta(\mathbf{r}).$$

The spin-polarization due to the exchange magnetic field can be found using linear response theory (Doniach and Sondheimer 1998b)

$$\mathbf{s}(\mathbf{r}) = \bar{J}_{sd} \int \chi_s(\mathbf{r} - \mathbf{r}', t - t')\, \mathbf{S}_{\mathrm{d}} \delta(\mathbf{r}) = \bar{J}_{sd} \chi_s(r) \mathbf{S}_{\mathrm{d}}.$$

χ_s is spin susceptibility

$$\chi_s(\mathbf{r}-\mathbf{r}',t-t') = \frac{i}{\hbar}\theta(t-t')\langle[s(\mathbf{r},t),s(\mathbf{r}',t')]\rangle$$

of a free electronic system. Its Fourier transform

$$\chi_s(\mathbf{q},\omega) = \lim_{\eta\to 0}\sum_{\mathbf{k}} \frac{n_{\mathbf{k}} - n_{\mathbf{k+q}}}{E_{\mathbf{k+q}} - E_{\mathbf{k}} + \hbar\omega + i\eta},$$

has a form similar to the charge susceptibility. However, in the Stoner model, $\chi_s(\mathbf{q},\omega)$ of interacting electron has been shown to be enhanced by positive U_s (Kim 1999b)

$$\chi_s(q=0,0) = \frac{\chi_s^0}{1-U_sN_F}.$$

see Appendix for derivation.

The inverse Fourier transform has spatial oscillation similar to the Friedel Oscillation (Mitsui et al. 2020)

$$\chi_s(r) = \sum_{\mathbf{q}} e^{i\mathbf{q}\cdot\mathbf{r}}\chi_s(q,\omega\to 0) = \frac{N_F}{8\pi r^3}\left(\frac{\sin 2k_F r}{2k_F r} - \cos 2k_F r\right).$$

$\chi_s(r)$ is well studied in RKKY interaction (Ruderman and Kittel 1954; Kasuya 1956; Yosida 1957). In the RKKY interaction, the spin-polarized conduction electrons mediate an exchange interaction between two localized spins. The strength of the indirect exchange strength determined by $\chi_s(r)$

$$J_{RKKY} = J_{sd}^2\chi_s(|\mathbf{r}_1 - \mathbf{r}_2|)$$

Trigonometric function in $\chi_s(r)$ indicates that the coupling can be ferromagnetic or antiferromagnetic depending on the distance.

The study of RKKY interaction shows that the indirect interaction also occurs at a magnetic multilayer (Stiles 1999; B. A. Jones and Hanna 1993). The spin polarization and coupling strength now depends on the thickness x of the nonmagnetic spacer between the magnetic layers (Bruno 1994)

$$J_{RKKY}^{multilayer} = \bar{J}_{sd}^{2} \chi_s(x),$$

$$\chi_s(x) = \frac{N_d}{A} \sum_q e^{iqx} \chi_s(q) \propto \left(\frac{\pi}{2} - \mathrm{Si}(2k_F x) - \frac{\cos 2k_F x}{2k_F x} + \frac{\sin 2k_F x}{(2k_F x)^2} \right)$$

here $\mathrm{Si}(x)$ is the sine integral function. $J_{RKKY}^{multilayer}$
J_{RKKY} of the pure 1D system (Cahaya 2022) because it has no contributions from the k-values that are not normal to the xy plane (Yafet 1987).

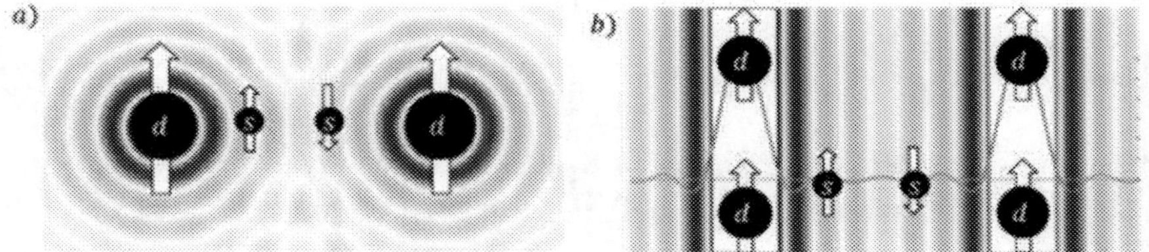

Figure 3. Ruderman-Kittel-Kasuya-Yosida-type indirect exchange interaction (a) between spins in the bulk and (b) between nearby magnetic layers mediated by conduction electrons.

4.2. Intralayer Dzyaloshinskii-Moriya Interaction

Besides s-d exchange interaction, the Coulomb interaction also affects the conduction electron via magnetic field

$$\mathbf{B} = \mu\varepsilon\mathbf{v} \times \mathbf{E} = -\mu\mathbf{v} \times \mathbf{\nabla}\left(\frac{1}{4\pi r}\right) = \frac{\mu}{4\pi r^2}\mathbf{v} \times \mathbf{r}$$

acting on electron spin $\mathbf{s}$. The magnetic field induces a spin-orbit coupling Hamiltonian on the conduction electron

$$H_{spin-orbit} = \frac{e}{m_e}\mathbf{s} \cdot \mathbf{B} = \frac{e\mu}{4\pi r^2 m_e^2}\mathbf{s} \cdot \mathbf{l} \equiv \Lambda(r)\mathbf{s} \cdot \mathbf{l},$$

$\mu = \mu_0 \mu_r$ is magnetic permeability. The relative permeability

$$\mu_r = 1 + \chi_m,$$

depends on the magnetic susceptibility $\chi_m \propto \chi_s$, which is enhanced by electron - electron interaction parameter U_s

$$\chi_s = \frac{\chi_s^0}{1 - UN_F}.$$

Therefore $\Lambda \propto 1/(1 - UN_F)$. The interplay between exchange and spin-orbit interactions is shown to be responsible for a Dzyaloshinskii–Moriya-type antisymmetric exchange interaction between two spins $\mathbf{S}_1$ and $\mathbf{S}_2$ of d-electrons (Fert and Levy 1980).

$$H_{DM} = -D(\mathbf{R}_1, \mathbf{R}_2)(\mathbf{S}_1 \times \mathbf{S}_2),$$

where $\mathbf{R}_1$ and $\mathbf{R}_2$ are positions of $\mathbf{S}_1$ and $\mathbf{S}_2$, respectively. While the original theory is for transition metal bulk, DM interaction has been observed in the bilayer of ferromagnet and nonmagnetic metal (Tacchi et al. 2017; Yang et al. 2015).

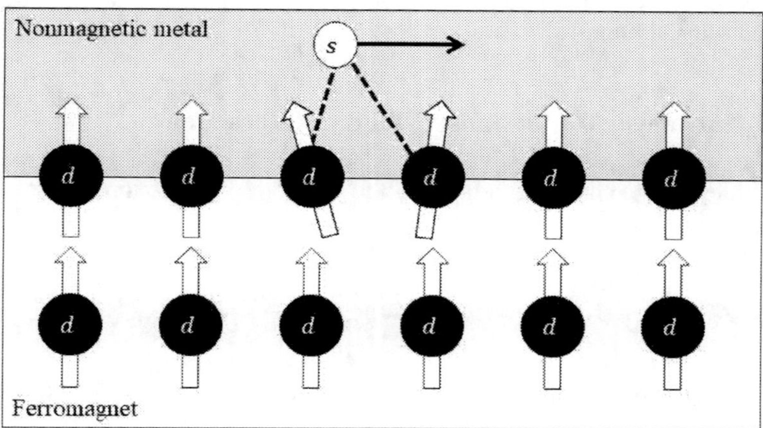

Figure 4. Dzyaloshinskii-Moriya-type Antisymmetric exchange interactions at magnetic interface due to interfacial spin-orbit interaction.

The exchange constant is directly proportional to J_{sd} Λ (Fert and Levy 1980).

$$D \propto J_{sd}^2 \Lambda,$$

Since $J_{sd} \propto \lambda^{-2} \propto (1 + U_s N_F)$, we can summarize the relation of D and the electron - electron interaction parameter U_s

$$D = \frac{D_0}{\varepsilon_r^2}\mu_r = D_0(1 + U_s N_F)^2\left(1 + \frac{\chi_m^0}{1 - U_s N_F}\right),$$

The values of D at the interface of Co and various heavy metals has been measured experimentally (Ma et al. 2018). The values of $U_s N_F$ metals are summarized in Table 2. Figure 4 shows the agreement of our theory and the experimental data for $\chi_m^0 = 0.70 \pm 0.16$.

Table 2. Dzyaloshinskii-Moriya interaction constants of Co|heavy metal (HM) and electron - electron interaction parameter $U_s N_F$

HM elements	$U_s N_F$ **(Sigalas and Papaconsantopoulos 1994)**	$\lvert D \rvert$ **(10^{-15}/m) (Ma et al. 2018)**
Au	0.05	218
Ta	0.34	79
W	0.10	108
Ir	0.29	336
Pt	0.73* (* Sitorus et al. 2021)	1588

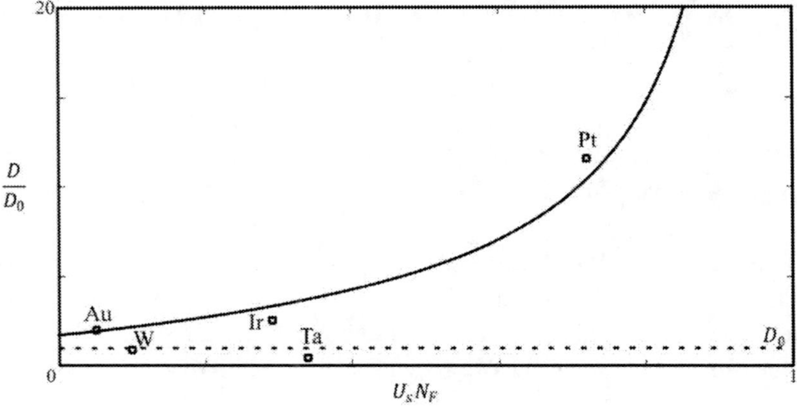

Figure 5. Antisymmetric exchange interaction of bilayer of Co and heavy metal. Values of D are normalized with D_0 (dashed line). The theoretical value from the screened exchange interaction (full line) agrees with the experimental data.

Conclusion

We discuss the exchange interaction at transition metal systems. The exchange interaction arises from Pauli exclusion principle of many electron systems. The exchange interaction, parameterized by J, depends on the electrostatic interaction between two electrons. In the Thomas-Fermi and Lindhard theories of metallic system, the screening of electric fields arises from the relative permittivity of the system. Due to the screening, the bare Coulomb interaction should be replaced by Yukawa-type screened Coulomb potential. We estimate the values J as a function of screening parameter.

The screening of Coulomb interaction also influences the exchange interaction between localized d-electrons and itinerant s-electrons. The s-d interaction is characterized by k_F/ζ and λ/ζ. Screening constant λ is influenced by the electron - electron interaction of the s-electron too. In the strong screening limit, the exchange interaction acts as a magnetic field that polarizes the spin of conduction electrons. The spin-polarization enables the conduction electrons to mediate the RKKY exchange interaction between two spins. The indirect RKKY interaction also occurs at magnetic multilayers.

When the s-d exchange interaction is accompanied by spin-orbit interaction, the conduction electron also mediates antisymmetric exchange interactions between d-electrons at the same interface. We show the agreement between the theoretical relation of Dzyaloshinskii-Moriya interaction constant to UN_F

Appendix: Charge and Spin Susceptibility of Interacting Electrons

Here we discuss the charge $\chi_c(\mathbf{q},\omega)$ and spin $\chi_s(\mathbf{q},\omega)$ susceptibilities of-interacting electrons, which were mentioned without proof in the maintext. To describe the susceptibilities of an interacting electrons system, we consider the following Hamiltonian (Doniach and Sondheimer 1998a)

$$H = \sum_{\mathbf{k}\sigma} E_{\mathbf{k}}\, a^{\dagger}_{\mathbf{k},\sigma} a_{\mathbf{k},\sigma} + \sum_{\mathbf{kpq}} U a^{\dagger}_{\mathbf{p+q},\uparrow} a_{\mathbf{p},\uparrow} a^{\dagger}_{\mathbf{k-q},\downarrow} a_{\mathbf{k},\downarrow}$$

$a_{k,\sigma}^{\dagger}$ and $a_{k,\sigma}$ are energy, creation and annihilation operators of conduction electron with wavenumber **k**, respectively. The first term is the free electronic Hamiltonian.

Charge Susceptibility

The charge susceptibility

$$\chi_c(\mathbf{r},t) = \frac{i}{\hbar}\theta(t)\langle[n(\mathbf{r},t), n(\mathbf{0},0)]\rangle$$

can be evaluated using the expression of charge density in second quantization

$$n(\mathbf{r},t) = \sum_{\mathbf{pq}\alpha} e^{i\mathbf{q}\cdot\mathbf{r}} a_{\mathbf{p+q},\alpha}^{\dagger}(t) a_{\mathbf{p},\alpha}(t).$$

θ is the Heaviside step function. Such that

$$\chi_c(\mathbf{r},t) = \frac{i\theta(t)}{\hbar}\sum_{\mathbf{pq}\alpha} e^{i\mathbf{q}\cdot\mathbf{r}} \langle[a_{\mathbf{p+q},\alpha}^{\dagger}(t) a_{\mathbf{p},\alpha}(t), n(\mathbf{0},0)]\rangle \equiv \sum_{\mathbf{pq}} e^{i\mathbf{q}\cdot\mathbf{r}} \chi_c(\mathbf{p},\mathbf{q},t)$$

The time derivation of $\chi_c(\mathbf{p},\mathbf{q},t)$ is

$$\frac{\partial \chi_c(\mathbf{p},\mathbf{q},t)}{\partial t} = \frac{i\delta(t)}{\hbar}\sum_{\alpha}\langle[a_{\mathbf{p+q},\alpha}^{\dagger} a_{\mathbf{p},\alpha}, n(\mathbf{0},0)]\rangle + \frac{\langle[\chi_c(\mathbf{p},\mathbf{q},t), H]\rangle}{i\hbar}$$

The first term on the right side can be approximated using random phase approximation[1]

$$\langle[a_{\mathbf{p+q},\alpha}^{\dagger} a_{\mathbf{p},\alpha}, n(\mathbf{0},0)]\rangle \simeq n_{\mathbf{p+q},\alpha} - n_{\mathbf{p},\alpha}$$

The second term can be evaluated using mean field approximation[2]

---

[1] $a_{p\alpha}^{\dagger} a_{q\beta} \simeq \delta_{pq}\delta_{\alpha\beta} n_{p\alpha}$.

[2] $a_{p\alpha}^{\dagger} a_{q\alpha} a_{k\beta}^{\dagger} a_{l\beta} \simeq n_{kl} a_{p\alpha}^{\dagger} a_{q\alpha} + n_{pq} a_{k\beta}^{\dagger} a_{l\beta}$.

$$[\chi_c(\mathbf{p},\mathbf{q},t),H] = \left(E_{\mathbf{p}} - E_{\mathbf{p+q}}\right)\chi_c(\mathbf{p},\mathbf{q},t) + U\sum_{\alpha}\left(n_{\mathbf{p+q},\alpha} - n_{\mathbf{p},\alpha}\right)\sum_{k}\chi_c(\mathbf{k},\mathbf{q},t).$$

One can arrive at the equation for $\chi_c(\mathbf{p},\mathbf{q},t)$

$$\left[i\hbar\frac{\partial}{\partial t} + E_{\mathbf{p+q}} - E_{\mathbf{p}}\right]\chi_c(\mathbf{p},\mathbf{q},t) = \sum_{\alpha}\left(n_{\mathbf{p},\alpha} - n_{\mathbf{p+q},\alpha}\right)\left[\delta(t) - U\sum_{k}\chi_c(\mathbf{k},\mathbf{q},t)\right]$$

that can be solved using its Fourier transform in frequency domain

$$\chi_c(\mathbf{p},\mathbf{q},\omega) = -\frac{\sum_{\alpha}\left(n_{\mathbf{p},\alpha} - n_{\mathbf{p+q},\alpha}\right)}{E_{\mathbf{p+q}} - E_{\mathbf{p}} + i\hbar\omega}\left[1 - U\sum_{k}\chi_c(\mathbf{k},\mathbf{q},\omega)\right]$$

$\chi_c(\mathbf{q},\omega)$ can be found by taking the sum of $\chi_c(\mathbf{p},\mathbf{q},\omega)$ over $\mathbf{p}$

$$\chi_c(\mathbf{q},\omega) = \frac{\chi_c^0(\mathbf{q},\omega)}{1 + U\chi_c^0(\mathbf{q},\omega)},$$

where

$$\chi_c^0(\mathbf{q},\omega) = \sum_{\mathbf{p}\alpha}\frac{n_{\mathbf{p},\alpha} - n_{\mathbf{p+q},\alpha}}{E_{\mathbf{p+q}} - E_{\mathbf{p}} + i\hbar\omega}.$$

The static limit of $\chi_c^0(\mathbf{q},\omega \to 0)$ can be evaluated using the low temperature limit of Fermi-Dirac distribution $n_{\mathbf{k},\alpha} = \theta(k_F - k)$

$$\begin{aligned}\chi_c^0(\mathbf{q},0) &= \sum_{\mathbf{p}\alpha} n_{\mathbf{p},\alpha}\left(\frac{1}{E_{\mathbf{p+q}} - E_{\mathbf{p}}} - \frac{1}{E_{\mathbf{p+q}} - E_{\mathbf{p-q}}}\right) \\ &= N_F\left(\frac{1}{2} + \frac{1 - (q/2k_F)^2}{2q/k_F}\ln\left|\frac{q + 2k_F}{q - 2k_F}\right|\right)\end{aligned}$$

Therefore, the long wavelength limit for the static charge susceptibility is reduced by U

$$\chi_c^0(\mathbf{0},0) = \frac{UN_F}{1+UN_F}.$$

Spin Susceptibility

The spin susceptibility depends on the spin pair

$$\chi_s^{ab}(\mathbf{r},t) = \frac{i}{\hbar}\theta(t)\langle[s_a(\mathbf{r},t), s_b(\mathbf{0},0)]\rangle$$

can be evaluated using the expression of spin density in second quantization

$$s_b(\mathbf{r},t) = \sum_{\mathbf{pq}\alpha\beta} e^{i\mathbf{q}\cdot\mathbf{r}} a^\dagger_{\mathbf{p}+\mathbf{q},\alpha}(t)\sigma^b_{\alpha\beta} a_{\mathbf{p},\beta}(t),$$

where $\sigma^b_{\alpha\beta}$ is the component of a Pauli matrix σ_b

$$(\sigma_x,\ \sigma_y,\ \sigma_z) = \left(\begin{pmatrix}0 & 1\\ 1 & 0\end{pmatrix}, \begin{pmatrix}0 & -i\\ i & 0\end{pmatrix}, \begin{pmatrix}1 & 0\\ 0 & -1\end{pmatrix}\right)$$

The Pauli matrices satisfy $\sigma_a\sigma_b = 1\delta_{ab} + i\epsilon_{abc}\sigma_c$. Such that

$$\begin{aligned}\chi_s^{ab}(\mathbf{r},t) &= \frac{i\theta(t)}{\hbar}\sum_{\mathbf{pq}\alpha\beta} e^{i\mathbf{q}\cdot\mathbf{r}}\left\langle\left[a^\dagger_{\mathbf{p}+\mathbf{q},\alpha}(t)\sigma^b_{\alpha\beta} a_{\mathbf{p},\beta}(t), s_b(\mathbf{0},0)\right]\right\rangle \\ &\equiv \sum_{\mathbf{pq}} e^{i\mathbf{q}\cdot\mathbf{r}}\,\chi_s^{ab}(\mathbf{p},\mathbf{q},t)\end{aligned}$$

For isotropic system conveniences, $\chi_s^{ab} = \delta_{ab}\chi_s$. Therefore, we only need to evaluate χ_s^{zz}

$$\frac{\partial\chi_s^{zz}(\mathbf{p},\mathbf{q},t)}{\partial t} = \frac{i\delta(t)}{\hbar}\sum_{\alpha\beta}\left\langle\left[a^\dagger_{\mathbf{p}+\mathbf{q},\alpha}(t)\sigma^z_{\alpha\beta} a_{\mathbf{p},\beta}(t), s_z(\mathbf{0},0)\right]\right\rangle + \frac{\langle[\chi_s^{zz}(\mathbf{p},\mathbf{q},t), H]\rangle}{i\hbar}$$

The first term on the right side can be evaluated as

$$\langle[a^{\dagger}_{\mathbf{p+q},\alpha}(t)\sigma^{z}_{\alpha\beta}a_{\mathbf{p},\beta}(t),s_b(\mathbf{0},0)]\rangle \simeq \delta_{\alpha\beta}\left(n_{\mathbf{p+q},\alpha}-n_{\mathbf{p},\alpha}\right).$$

On the other hand, the second term can be evaluated as

$$\begin{aligned}[\chi_s^{zz}(\mathbf{p},\mathbf{q},t),H] &= \left(E_{\mathbf{p}}-E_{\mathbf{p+q}}\right)\chi_s^{zz}(\mathbf{p},\mathbf{q},t)\\ &\quad + U\sum_{\alpha}\left(n_{\mathbf{p},\alpha}-n_{\mathbf{p+q},\alpha}\right)\sum_{k}\chi_s^{zz}(\mathbf{k},\mathbf{q},t)\end{aligned}$$

One can arrive at the equation for $\chi_c(\mathbf{p},\mathbf{q},t)$

$$\begin{aligned}&\left[i\hbar\frac{\partial}{\partial t}+E_{\mathbf{p+q}}-E_{\mathbf{p}}\right]\chi_s^{zz}(\mathbf{p},\mathbf{q},t)\\ &\qquad = \sum_{\alpha}\left(n_{\mathbf{p},\alpha}-n_{\mathbf{p+q},\alpha}\right)\left[\delta(t)+U\sum_{k}\chi_s^{zz}(\mathbf{k},\mathbf{q},t)\right]\end{aligned}$$

that can be solved using its Fourier transform in frequency domain

$$\chi_s^{zz}(\mathbf{p},\mathbf{q},\omega) = -\frac{\sum_{\alpha}\left(n_{\mathbf{p},\alpha}-n_{\mathbf{p+q},\alpha}\right)}{E_{\mathbf{p+q}}-E_{\mathbf{p}}+i\hbar\omega}\left[1+U\sum_{k}\chi_s^{zz}(\mathbf{k},\mathbf{q},\omega)\right]$$

$\chi_s^{zz}(\mathbf{q},\omega)$ can be found by taking the sum of $\chi_c(\mathbf{p},\mathbf{q},\omega)$ over $\mathbf{p}$

$$\chi_s^{zz}(\mathbf{q},\omega) = \frac{\chi_s^{0}(\mathbf{q},\omega)}{1+U\chi_s^{0}(\mathbf{q},\omega)}$$

where

$$\chi_s^{0}(\mathbf{q},\omega) = \sum_{\mathbf{p}\alpha}\frac{n_{\mathbf{p},\alpha}-n_{\mathbf{p+q},\alpha}}{E_{\mathbf{p+q}}-E_{\mathbf{p}}+i\hbar\omega}.$$

The static limit of $\chi_s^{0}(\mathbf{q},\omega\to 0)$ can be evaluated

$$\chi_s^{0}(\mathbf{q},0) = \sum_{\mathbf{p}\alpha} n_{\mathbf{p},\alpha}\left(\frac{1}{E_{\mathbf{p+q}}-E_{\mathbf{p}}}-\frac{1}{E_{\mathbf{p+q}}-E_{\mathbf{p-q}}}\right)$$

$$= N_F \left(\frac{1}{2} + \frac{1 - (q/2k_F)^2}{2q/k_F} \ln \left| \frac{q + 2k_F}{q - 2k_F} \right| \right)$$

Therefore, the long wavelength limit for the spin charge susceptibility is increased by U

$$\chi_s^0(\mathbf{0}, 0) = \frac{UN_F}{1 - UN_F}.$$

References

Andrade-Neto, A. V. 2016. "Dielectric Function for Free Electron Gas: Comparison between Drude and Lindhard Models." *Revista Brasileira de Ensino de Física* [Brazilian Journal of Physics Teaching] 39 (2). https://doi.org/10.1590/1806-9126-rbef-2016-0206.

Barnes, S. E. 2004. "Theory of Superconductivity." In *Physics of Transition Metal Oxides*, edited by Sadamichi Maekawa, Takami Tohyama, Stewart E. Barnes, Sumio Ishihara, Wataru Koshibae, and Giniyat Khaliullin, 101-66. Berlin, Heidelberg: Springer Berlin Heidelberg. https://doi.org/10.1007/978-3-662-09298-9_3.

Bena, Cristina. 2016. "Friedel Oscillations: Decoding the Hidden Physics." *Comptes Rendus Physique* [Physical Reports] 17 (3): 302-21. https://doi.org/https://doi.org/10.1016/j.crhy.2015.11.006.

Bruno, P. 1994. "Comment on ("Contribution of Quantum-Well States to the RKKY Coupling in Magnetic Multilayers")" *Physical Review Letters* 72 (22): 3627. https://doi.org/10.1103/PhysRevLett.72.3627.

Burke, Kieron. 2012. "Perspective on Density Functional Theory." *The Journal of Chemical Physics* 136 (15): 150901. https://doi.org/10.1063/1.4704546.

Cahaya, Adam B. 2018. "Spin, Charge, and Heat Coupling at Magnetic Interfaces." https://ci.nii.ac.jp/naid/500001078235.

———. 2022. "Adiabatic Limit of RKKY Range Function in One Dimension." *Journal of Magnetism and Magnetic Materials* 547: 168874. https://doi.org/10.1016/j.jmmm.2021.168874.

Cahaya, Adam B., Anugrah Azhar, and Muhammad Aziz Majidi. 2021. "Yukawa Potential for Realistic Prediction of Hubbard and Hund Interaction Parameters for Transition Metals." *Physica B: Condensed Matter* 604: 412696. https://doi.org/10.1016/j.physb.2020.412696.

Cahaya, Adam B., and Muhammad Aziz Majidi. 2021. "Effects of Screened Coulomb Interaction on Spin Transfer Torque." *Physical Review B* 103 (9): 094420. https://doi.org/10.1103/PhysRevB.103.094420.

Doniach, S., and E. H. Sondheimer. 1998a. "The Interacting Electron Gas." In *Green's Functions for Solid State Physicists*, 123-56. Published by Imperial College Press and Distributed by World Scientific Publishing Co. https://doi.org/doi:10.1142/9781860944031_0006.

———. 1998b. "The Magnetic Instability of the Interacting Electron Gas." In *Green's Functions for Solid State Physicists*, 157-83. Published by Imperial College Press and Distributed by World Scientific Publishing Co. https://doi.org/doi:10.1142/9781860944031_0007.

Eder, Robert. 2012. "Multiplets in Transition Metal Ions." *Correlated Electrons: From Models to Materials. Modeling and Simulation*; Pavarini, E., Koch, E., Anders, F., Jarrell, M., Eds.

Fernández, Francisco M. 2008. "Comment on: 'Series Solution to the Thomas-Fermi Equation' [Phys. Lett. A 365 (2007) 111]." *Physics Letters A* 372 (31): 5258-60. https://doi.org/https://doi.org/10.1016/j.physleta.2008.05.071.

Fert, A., and Peter M. Levy. 1980. "Role of Anisotropic Exchange Interactions in Determining the Properties of Spin-Glasses." *Physical Review Letters* 44 (23): 1538-41. https://doi.org/10.1103/PhysRevLett.44.1538.

Georges, Antoine, Luca de' Medici, and Jernej Mravlje. 2013. "Strong Correlations from Hund's Coupling." *Annual Review of Condensed Matter Physics* 4 (1): 137-78. https://doi.org/10.1146/annurev-conmatphys-020911-125045.

Heisenberg, W. 1928. "Zur Theorie Des Ferromagnetismus [On the theory of ferromagnetism]." *Zeitschrift Für Physik* [Magazine for physics] 49 (9): 619-36. https://doi.org/10.1007/BF01328601.

Imada, Masatoshi, Atsushi Fujimori, and Yoshinori Tokura. 1998. "Metal-Insulator Transitions." *Reviews of Modern Physics* 70 (4): 1039-1263. https://doi.org/10.1103/RevModPhys.70.1039.

Isihara, Akira. 1998. "Dielectric Function." In *Electron Liquids*, edited by Akira Isihara, 21-40. Berlin, Heidelberg: Springer Berlin Heidelberg. https://doi.org/10.1007/978-3-642-80392-5_2.

Jones, Barbara A., and C. B. Hanna. 1993. "Contribution of Quantum-Well States to the RKKY Coupling in Magnetic Multilayers." *Physical Review Letters* 71 (25): 4253-56. https://doi.org/10.1103/PhysRevLett.71.4253.

Jones, R. O., and O. Gunnarsson. 1989. "The Density Functional Formalism, Its Applications and Prospects." *Reviews of Modern Physics* 61 (3): 689-746. https://doi.org/10.1103/RevModPhys.61.689.

Kasuya, T. 1956. "A Theory of Metallic Ferro- and Antiferromagnetism on Zener's Model." *Prog. Theor. Phys.* 16: 45.

Kim, Duk Joo. 1999a. "Linear Responses of Metallic Electrons." In *New Perspectives in Magnetism of Metals*, edited by Duk Joo Kim, 85-146. Boston, MA: Springer US. https://doi.org/10.1007/978-1-4757-3052-4_4.

———. 1999b. "Mean Field Theory of Magnetism of Metals: The Stoner Model." In *New Perspectives in Magnetism of Metals*, edited by Duk Joo Kim, 57-83. Boston, MA: Springer US. https://doi.org/10.1007/978-1-4757-3052-4_3.

King, Rollin A., and Nicholas C. Handy. 2000. "Kinetic Energy Functionals from the Kohn-Sham Potential." *Phys. Chem. Chem. Phys.* 2 (22): 5049-56. https://doi.org/10.1039/b005896n.

Kohn, W. 1999. "Nobel Lecture: Electronic Structure of Matter - Wave Functions and Density Functionals." *Reviews of Modern Physics* 71 (5): 1253-66. https://doi.org/10.1103/RevModPhys.71.1253.

Kondo, J. 1962. "Anomalous Hall Effect and Magnetoresistance of Ferromagnetic Metals." *Prog. Theor. Phys.* 27 (4): 772-92. https://doi.org/10.1143/PTP.27.772.

Lieb, Elliott H., and Barry Simon. 1977. "The Thomas-Fermi Theory of Atoms, Molecules and Solids." *Advances in Mathematics* 23 (1): 22-116. https://doi.org/10.1016/0001-8708(77)90108-6.

Lindhard, J. 1954. "On the Properties of a Gas of Charged Particles." *Dan. Vid. Selsk Mat.-Fys. Medd.* 28: 8. https://cir.nii.ac.jp/crid/1573105974901993344.

Ma, Xin, Guoqiang Yu, Chi Tang, Xiang Li, Congli He, Jing Shi, Kang L Wang, and Xiaoqin Li. 2018. "Interfacial Dzyaloshinskii-Moriya Interaction: Effect of $5d$ Band Filling and Correlation with Spin Mixing Conductance." *Physical Review Letters* 120 (15): 157204. https://doi.org/10.1103/PhysRevLett.120.157204.

March, N. H. 1983. "Origins - The Thomas-Fermi Theory." In *Theory of the Inhomogeneous Electron Gas*, edited by S. Lundqvist and N. H. March, 1-77. Boston, MA: Springer US. https://doi.org/10.1007/978-1-4899-0415-7_1.

Miraglia, J. E., and M. S. Gravielle. 2002. "Surface Dielectric Functions of a Free-Electron Gas." *Physical Review A* 66 (3): 32901. https://doi.org/10.1103/PhysRevA.66.032901.

Mitsui, T., S. Sakai, S. Li, T. Ueno, T. Watanuki, Y. Kobayashi, R. Masuda, M. Seto, and H. Akai. 2020. "Magnetic Friedel Oscillation at the Fe(001) Surface: Direct Observation by Atomic-Layer-Resolved Synchrotron Radiation Fe 57 Mössbauer Spectroscopy." *Physical Review Letters* 125 (23): 236806. https://doi.org/10.1103/PhysRevLett.125.236806.

Morgan III, John. 2006. "Thomas-Fermi and Other Density-Functional Theories." In *Springer Handbook of Atomic, Molecular, and Optical Physics*, edited by Gordon Drake, 295-306. New York, NY: Springer New York. https://doi.org/10.1007/978-0-387-26308-3_20.

Oles, A. M., and G. Stollhoff. 1984. "Correlation Effects in Ferromagnetism of Transition Metals." *Physical Review B* 29 (1): 314-27. https://doi.org/10.1103/PhysRevB.29.314.

Pauling, Linus Carl, and Arnold Johannes Wilhelm Sommerfeld. 1927. "The Theoretical Prediction of the Physical Properties of Many Electron Atoms and Ions. Mole Refraction, Diamagnetic Susceptibility, and Extension in Space." *Proceedings of the Royal Society of London. Series A, Containing Papers of a Mathematical and Physical Character* 114 (767): 181-211. https://doi.org/10.1098/rspa.1927.0035.

Plindov, G. I., and S. K. Pogrebnya. 1987. "The Analytical Solution of the Thomas-Fermi Equation for a Neutral Atom." *Journal of Physics B: Atomic and Molecular Physics* 20 (17): L547-50. https://doi.org/10.1088/0022-3700/20/17/001.

Pople, John A. 1999. "Nobel Lecture: Quantum Chemical Models." *Reviews of Modern Physics* 71 (5): 1267-74. https://doi.org/10.1103/RevModPhys.71.1267.

Resta, Raffaele. 1977. "Thomas-Fermi Dielectric Screening in Semiconductors." *Physical Review B* 16 (6): 2717-22. https://doi.org/10.1103/PhysRevB.16.2717.

Ruderman, M. A., and C. Kittel. 1954. "Indirect Exchange Coupling of Nuclear Magnetic Moments by Conduction Electrons." *Phys. Rev.* 96 (1): 99-102.

Sigalas, M. M., and D. A. Papaconstantopoulos. 1994. "Calculations of the Total Energy, Electron-Phonon Interaction, and Stoner Parameter for Metals." *Physical Review B* 50 (11): 7255-61. https://doi.org/10.1103/PhysRevB.50.7255.

Sitorus, R. M., A. Azhar, A. B. Cahaya, A. R. T. Nugraha, and M. A. Majidi. 2021. "Theoretical Study of Complex Susceptibility of Pd and Pt." *Journal of Physics: Conference Series* 1816 (1): 012044. https://doi.org/10.1088/1742-6596/1816/1/012044.

Slater, J. C. 1929. "The Theory of Complex Spectra." *Physical Review* 34 (10): 1293-1322. https://doi.org/10.1103/PhysRev.34.1293.

———. 1930. "Atomic Shielding Constants." *Physical Review* 36 (1): 57-64. https://doi.org/10.1103/PhysRev.36.57.

Solovej, Jan Philip. 2016. "A New Look at Thomas-Fermi Theory." *Molecular Physics* 114 (7-8): 1036-40. https://doi.org/10.1080/00268976.2015.1130273.

Stiles, M D. 1999. "Interlayer Exchange Coupling." *Journal of Magnetism and Magnetic Materials* 200 (1): 322-37. https://doi.org/https://doi.org/10.1016/S0304-8853(99)00334-0.

Tacchi, S., R. E. Troncoso, M. Ahlberg, G. Gubbiotti, M. Madami, J. Åkerman, and P. Landeros. 2017. "Interfacial Dzyaloshinskii-Moriya Interaction in $\mathrm{Pt}/\mathrm{CoFeB}$ Films: Effect of the Heavy-Metal Thickness." *Physical Review Letters* 118 (14): 147201. https://doi.org/10.1103/PhysRevLett.118.147201.

Vaugier, Loïg, Hong Jiang, and Silke Biermann. 2012. "Hubbard U and Hund Exchange J in Transition Metal Oxides: Screening versus Localization Trends from Constrained Random Phase Approximation." *Physical Review B* 86 (16): 165105. https://doi.org/10.1103/PhysRevB.86.165105.

Vleck, J. H. van. 1932. "Theory of the Variations in Paramagnetic Anisotropy among Different Salts of the Iron Group." *Physical Review* 41 (2): 208-15. https://doi.org/10.1103/PhysRev.41.208.

Wang, Yong, Xiaohui Liu, J. D. Burton, Sitaram S. Jaswal, and Evgeny Y. Tsymbal. 2012. "Ferroelectric Instability Under Screened Coulomb Interactions." *Physical Review Letters* 109 (24): 247601. https://doi.org/10.1103/PhysRevLett.109.247601.

Yafet, Y. 1987. "Ruderman-Kittel-Kasuya-Yosida Range Function of a One-Dimensional Free-Electron Gas." *Phys. Rev. B* 36 (7): 3948-49.

Yang, Hongxin, André Thiaville, Stanislas Rohart, Albert Fert, and Mairbek Chshiev. 2015. "Anatomy of Dzyaloshinskii-Moriya Interaction at Co/Pt Interfaces." *Physical Review Letters* 115 (26): 267210. https://doi.org/10.1103/PhysRevLett.115.267210.

Yosida, K. 1957. "Magnetic Properties of Cu-Mn Alloys." *Phys. Rev.* 106 (5): 893-98. https://doi.org/10.1103/PhysRev.106.893.

Biographical Sketch

Adam B. Cahaya

Affiliation: Department of Physics, Faculty of Mathematic and Natural Sciences, Universitas Indonesia

Education:

- 2013: BSc from Tohoku University
- 2015: MSc from Tohoku University
- 2018: DrSc from Tohoku University

Business Address: Fakultas MIPA, Kampus UI Depok, Jawa Barat, Indonesia 16424

Research and Professional Experience:

Theoretical Condensed Matter Physics of Magnetism and Magnetic Materials, with focus on:

- Origin of magnetism in strongly correlated materials
- Indirect exchange interactions
- Spin pumping and spin transfer torque in magnetic multilayer
- Thermoelectrics with spin current
- Spin-orbit coupling in rare-earth systems

Professional Appointments:

- 2019-now: Assistant Professor in Department of Physics, Universitas Indonesia
- 2019: Visiting Assistant Professor in Institute for Materials Research, Tohoku University
- 2018-2019: Lecturer in Department of Physics, Universitas Indonesia
- 2015-2018: Japan Society for the Promotion of Science (JSPS) Fellowship for Young Scientists.

Honors:

- 2008 Satyalancana Wira Karya from President of Indonesia
- 2012 青葉理学振興会奨励賞 [Aoba Science Promotion Association Encouragement Award] from Aoba Society of Science

Publications from the Last 3 Years:

1. Cahaya, Adam B. 2022. "Enhancement of Thermal Spin Pumping by Orbital Angular Momentum of Rare Earth Iron Garnet." *Journal of Magnetism and Magnetic Materials.* Elsevier BV. https://doi.org/10.1016/j.jmmm.2022.169248.

2. Cahaya, Adam B., Rico M. Sitorus, Anugrah Azhar, Ahmad R. T. Nugraha, and Muhammad Aziz Majidi. 2022. "Enhancement of spin-mixing conductance by s-d orbital hybridization in heavy metals." *Physical Review B.* American Physical Society (APS). https://doi.org/10.1103/PhysRevB.105.214438.
3. Muntini, Melania S., Edi Suprayoga, Sasfan A. Wella, Iim Fatimah, Lila Yuwana, Tosawat Seetawan, Adam B. Cahaya, Ahmad R. T. Nugraha, and Eddwi H. Hasdeo. 2022. "Spin-tunable thermoelectric performance in monolayer chromium pnictides." *Physical Review Materials*. American Physical Society (APS). https://doi.org/10.1103/PhysRevMaterials.6.064010.
4. Cahaya, Adam B., Anugrah Azhar, Dede Djuhana, and Muhammad Aziz Majidi. 2022. "Effect of Interfacial Spin Mixing Conductance on Gyromagnetic Ratio of Gd Substituted Y3Fe5O12." *Physics Letters A*. Elsevier BV. https://doi.org/10.1016/j.physleta.2022. 128085.
5. Cahaya, Adam B. 2022. "Adiabatic Limit of RKKY Range Function in One Dimension." *Journal of Magnetism and Magnetic Materials*. Elsevier BV. https://doi.org/10.1016/j.jmmm.2021.168874.
6. Cahaya, Adam B. 2021. "Antiferromagnetic Spin Pumping via Hyperfine Interaction." *Hyperfine Interactions*. Springer Science and Business Media LLC. https://doi.org/10.1007/s10751-021-01780-0.
7. Cahaya, Adam B., and Muhammad Aziz Majidi. 2021. "Effects of Screened Coulomb Interaction on Spin Transfer Torque." *Physical Review B.* American Physical Society (APS). https://doi.org/10.1103/physrevb.103.094420.
8. Cahaya, Adam B., Anugrah Azhar, and Muhammad Aziz Majidi. 2021. "Yukawa Potential for Realistic Prediction of Hubbard and Hund Interaction Parameters for Transition Metals." *Physica B: Condensed Matter.* Elsevier BV. https://doi.org/10.1016/j.physb. 2020.412696.
9. Cahaya, Adam B., Alejandro O. Leon, Mojtaba Rahimi Aliabad, and Gerrit E. W. Bauer. 2021. "Equilibrium Current Vortices in Simple Metals Doped with Rare Earths." *Physical Review B*. American Physical Society (APS). https://doi.org/10.1103/physrevb.103.064433.
10. Rangkuti, Choirun Nisaa, Adam B. Cahaya, Anugrah Azhar, Muhammad Aziz Majidi, and Andrivo Rusydi. 2019. "Manifestation of Charge/Orbital Order and Charge Transfer in Temperature-Dependent Optical Conductivity of Single-Layered Pr0.5Ca1.5MnO4." *Journal of*

Physics: Condensed Matter. IOP Publishing. https://doi.org/10.1088/1361-648x/ab2433.

11. Leon, Alejandro O., Jose d'Albuquerque e Castro, Juan C. Retamal, Adam B. Cahaya, and Dora Altbir. 2019. "Manipulation of the RKKY Exchange by Voltages." *Physical Review B*. American Physical Society (APS). https://doi.org/10.1103/physrevb.100.014403.

Chapter 4

Synthesis, Characterisation and Use of the Copper(II)-1,4-Bis-(3-Hexafluoroacetylacetonate) Benzene Framework

Palani Natarajan*, **PhD**
Department of Chemistry and Centre for Advanced Studies in Chemistry,
Panjab University, Chandigarh, India

Abstract

At ambient temperature, a new fluorinated β-diketone ligand, 1,4-bis-(3-hexafluoroacetylacetonate)benzene (L), was synthesised and utilised to build a metal-organic framework, ML, with Cu(II) ions. Both L and ML were found to be soluble in a wide variety of solvents including tetrahydrofuran, dichloromethane, chloroform, ethyl acetate, acetone, and isopropanol. Thus, the ML was extensively characterized by infrared spectroscopy, powder X-ray diffraction, cyclic voltammetry, thermal analysis, field emission scanning electron microscopy, energy dispersive spectroscopy, atomic force microscopy, and electron paramagnetic resonance spectroscopy. The 4-coordinated Cu(II) ions of ML have been shown to be useful for electrochemical caffeine detection in preliminary electrochemical investigations.

---

* Corresponding Author's Email: pnataraj@pu.ac.in.

In: Transition Metals - An Overview
Editor: Veronica J. Avila
ISBN: 979-8-88697-646-5
© 2023 Nova Science Publishers, Inc.

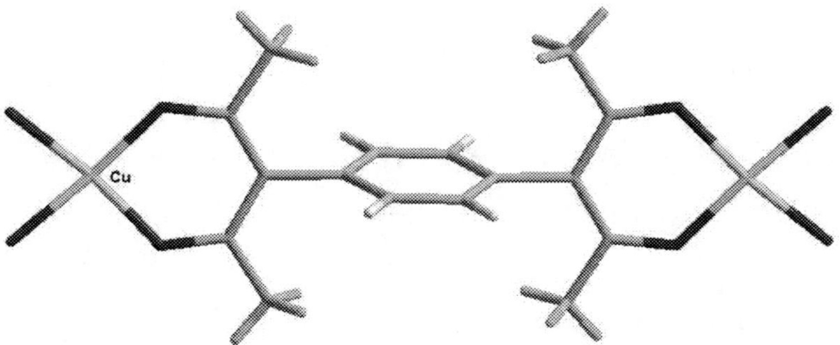

Introduction

Coordination polymers [1] constitute an important class of compounds in the exploratory research area of advanced materials design. In this connection, recently researchers have shown interest into the possibility of synthesizing fluorinated metal–organic frameworks [2] (FMOFs) because FMOFs tend to have exceptional chemical and biological inertness to the hydrogen analogues [1]. While compared to the C–H bond, the C–F bond is stronger (bond-dissociation energy, 130 ± 3 kcal/mol),[3] more polarized and possesses low-lying HOMO levels [4] that were demonstrated possible reasons for FMOFs greater constancy even at elevated temperatures. Accordingly, FMOFs are often being employed as zeolite-like material [1] for molecular selection, [4] adsorption, [5] ion exchange, [3] heterogeneous catalysis [2] and so on. However, the elaboration of FMOFs has been an ongoing challenge [2-5] because of limited solubility of these polymeric materials in common solvents, thus restricting the usages in sensor and catalysis fields [6].

The replacements of highly conjugated and rigid organic linkers of a coordination polymer by less conjugated and flexible analogues were proved to exhibit an improvement in the chemical solubility [7]. For example a semi-rigid ligand, benzene-1,4-dioxylacetate, leads to the soluble coordination networks when reacted with metal ions, whereas the rigid naphthalene-2,6-dioxylacetate linker yielded insoluble polymers [1]. Likewise the flexible bis-(3-acetylacetonates) ligands, tetraacetyl ethanes, [8] lead to soluble coordination polymers with metal ions of the first transition series and lanthanides. Nevertheless, very few publications dealing with the metal complexes of flexible- or semi-rigid ligands are available [7]. Moreover, the use of substituted hexafluoro-1,3-diketone linkers for creation of FMOFs has not been reported despite the fact that hexafluoro-*β*-diketones have shown unusual solubility and chelating ability with metal ions [9]. These findings fostered us to study the coordination properties of the novel 1,4-bis-(3-hexafluoroacetylacetonate)benzene (**L**, Scheme 1) toward Cu(II) ions. The complex, **ML**, obtained has been fully characterized by several techniques such as infrared spectroscopy (IR), powder X-ray diffraction (PXRD), cyclic voltammetry, thermogravimetric analysis, field emission scanning electron microscopy (FESEM), energy dispersive spectroscopy (EDS), atomic force microscopy (AFM) and electron paramagnetic resonance spectroscopy (EPR). In addition, we report the electrochemical sensing of caffeine by **ML** (Scheme 1) at ambient conditions. To our knowledge, this is the first soluble fluorinated polymeric network based on the *bis*-3-hexafluoro-*β*-diketonate linker reported until to date.

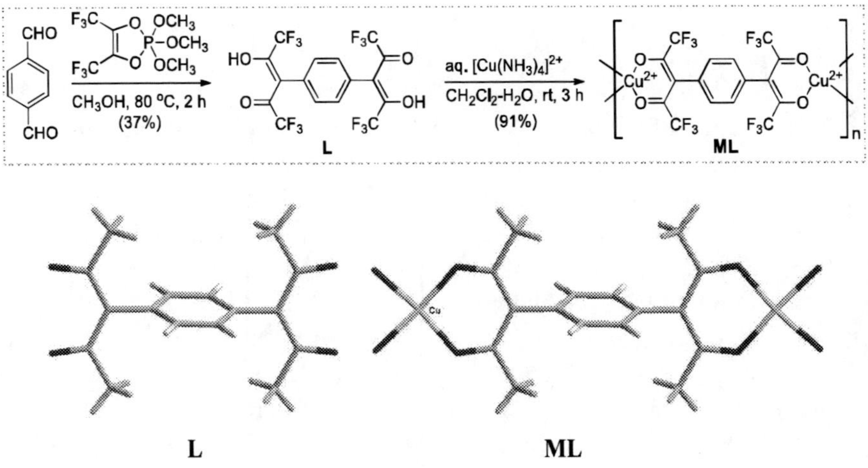

Scheme 1. A synthetic route to **L** and **ML** (top) and their simulated structures (bottom).

Experimental Section

General Information

All reactions were performed under a nitrogen gas atmosphere in oven-dried glassware. The solvents were freshly distilled prior to use. Raw materials were purchased from commercial source, and were used as received. The powder X-ray diffraction patterns were recorded on a Siemens D500 diffractometer with graphite monochromated and Cu Kα radiation (λ = 1.54056 Å). FT-IR spectra were recorded using a Perkin Elmer spectrum 100 instrument by using KBr pellets. Thermal analyses were performed in a Mettler DSC7 DSC under nitrogen gas atmosphere. NMR spectra were recorded on a Bruker Avance 400 MHz spectrometer in deuterated solvents. AFM images were done by a Nanoscope III Veeco (Digital Instrument, NY). Elemental analysis was measured using an Elementar Vario EL III. All measurements were carried out at ambient conditions unless stated otherwise.

Synthesis of L

To a solution of terephthalaldehyde (4.0 g, 30 mmol) in 20 ml of dry CH_2Cl_2 at 0 °C was slowly added 2,2,2-trimethoxyhexafluoromethyl-1,3-dioxaphospholene (13.9 g, 66 mmol) under nitrogen gas atmosphere. The resultant colorless solution was stirred at room temperature for 24 h. Subsequently, all volatiles were evaporated in vacuum and the residue was dissolved in 40 ml of anhydrous methanol and refluxed for 2 h. The solvent was evaporated to afford a yellow oil that was purified by column chromatography over silica gel using 20% ethyl acetate in n-hexane, followed by recrystallization from ethyl acetate and n-hexane gave ligand **L** as a colorless solid (yield 37%): ^{1}H NMR (400 MHz, $CDCl_3$) δ 16.8 (s, 2H, OH), 7.52 (s, 4H, Ar); ^{13}C NMR (100 MHz, $CDCl_3$) δ 192.6, 146.8, 132.2, 131.9, 119.7, 117.2; Anal. Calcd for $C_{16}H_6F_{12}O_4$: C, 39.20; H, 1.23. Found: C, 39.21; H, 1.21.

Synthesis of ML

To a solution of $CuSO_4 \cdot 5H_2O$ (0.35 g, 1.42 mmol) in water (5 mL) was added solution of concentrated aqueous ammonia until a clear dark blue solution formed. Then, the ligand (**L**, 0.42 g, 1.36 mmol) in dichloromethane (10 mL) was added slowly. Resultant two-phase solution was stirred under N_2 for 3 h. Subsequently, the reaction mixture was extracted with dichloromethane (2 × 100 mL) and dried over anhydrous Na_2SO_4 and evaporated. The pure product was obtained after recrystallization from methanol-dichloromethane solution furnishing **ML** as a blue solid (yield, 91%): IR (KBr, cm^{-1}) 3426, 2987, 2896, 2888, 1663, 1613, 1564, 1537, 1483, 1350, 1225, 1147, 1098, 808, 756, 729, 683, 666, 581.

Results and Discussion

The reaction of $Cu(NH_3)_4^{2+}$ with **L** in a mixture of dichloromethane-water solvents at room temperature gave **ML** (Scheme 1) in a yield of 91%, cf. Experimental Section. The stoichiometry and the purity of synthesized materials were confirmed by elemental analysis, IR, and EDS measurements. As **L** and **ML** were found to be stable to air and moisture, the measurements were conducted without any precaution unless stated otherwise.

Solubility of ML

We determined the solubility of **ML** in different solvents ranging from non-polar to polar medium. We found that the **ML** was readily soluble in toluene, chloroform, dichloromethane, isopropanol, acetone, THF and so on. However, it was insoluble in water and aqueous alcohols. Thanks to good solubility of complex in non-polar solvents, as allowed us to characterize **ML** by different techniques in a convenient manner. We noticed that the $CH_2Cl_2/CHCl_3$ solutions of **ML** show an olive-green color, whereas a blue color was observed in tetrahydrofuran/acetone/acetonitrile. This color change was probably due to the solvent coordination to the Cu(II) centers as similar behaviors were reported previously for several Cu(II) complexes [10].

PXRD of ML

To investigate the crystalline properties of **ML** in the solid state, X-ray diffraction was performed in the reflection mode. As shown in Figure 1, the solids of **ML** are primarily crystalline in nature. Since high intensity Bragg diffraction peaks are observed at 2θ = 10.34, 11.21, 14.48, 25.01, 28.98 and 32.97 in addition to a few low intensity peaks. The peak of highest intensity in diffractogram at 2θ = 14.48 is indexed as the d_{200} for Cu(II) ions [11]. Interestingly peaks for Cu_2O, a common byproduct of Cu(II) ions, were not detected indicating the high purity and stability of the **ML** examined [10].

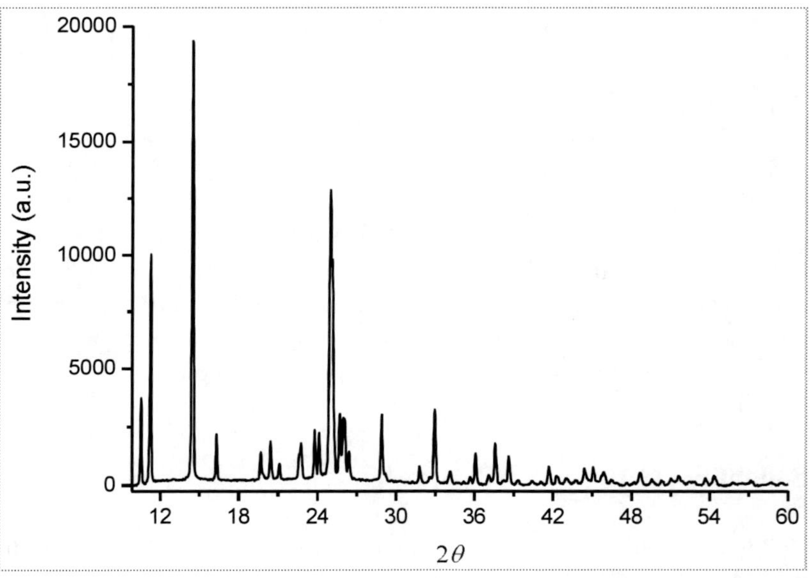

Figure 1. PXRD pattern of solids of the **ML** measured at ambient conditions.

FESEM of ML

The morphology of particles of the **ML** was studied by FESEM after drop casted on a metal stub. As shown in Figure 2, the **ML** has nonuniformly distributed eclipse shape flakes with the average diameter of 3.5 μm, which is similar to the value estimated from PXRD pattern (Figure 1) [11].

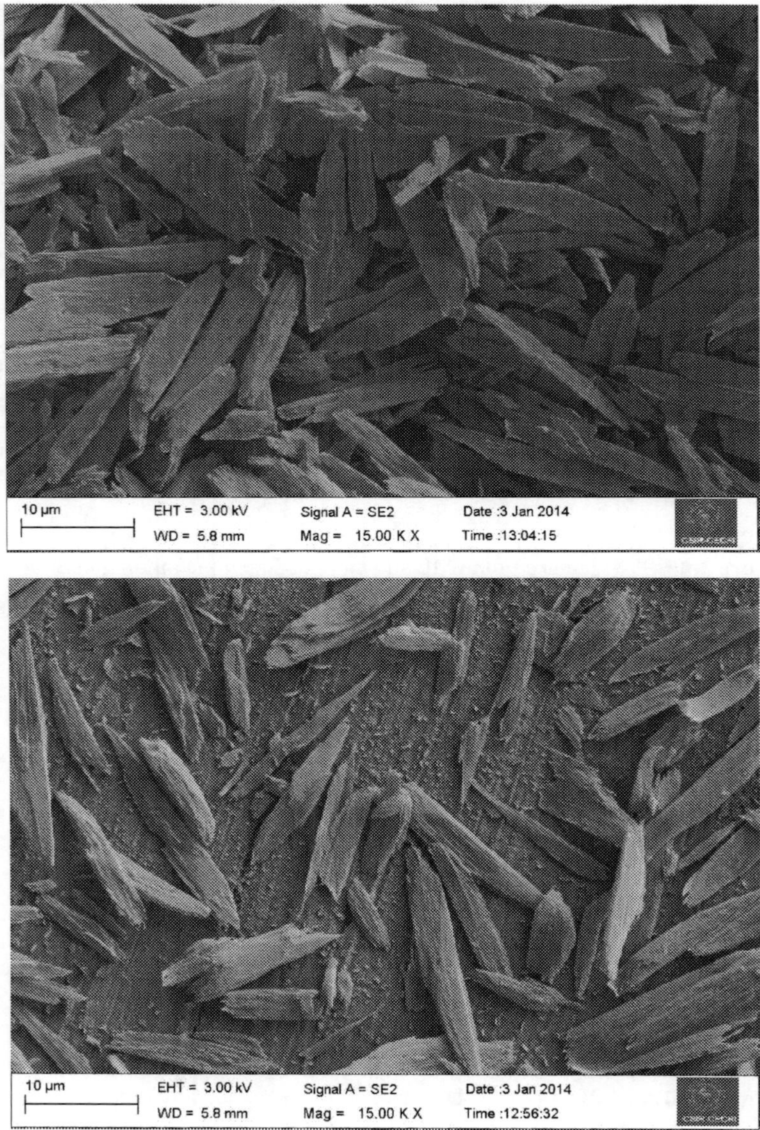

Figure 2. (Continued)

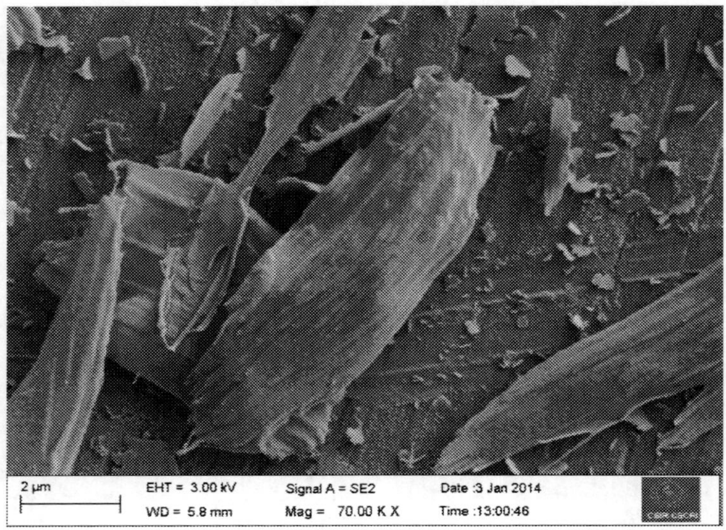

Figure 2. FESEM images of the ML shown in different magnifications.

EDS of ML

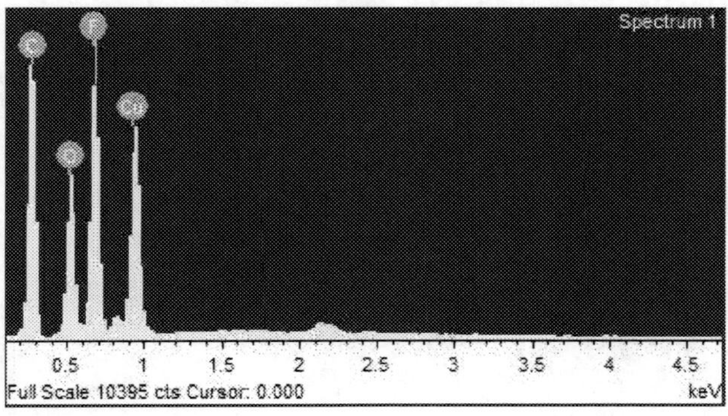

Figure 3. EDS analysis of the **ML** carried out at ambient conditions.

The chemical composition of **ML** was established by EDS measurements and the results are presented in Figure 3. Quantitative analyses of the EDS result show that the **ML** consists of 24.76% weight of carbon with atomic percentage of 25.56, and 18.82% weight of oxygen with atomic percentage of 14.44, and 45.19% weight of fluorine with atomic percentage of 48.13, and 11.23% weight of copper with atomic percentage of 11.87. Clearly, the EDS study confirms the presence of carbon, oxygen, fluorine and copper as the elements assembled in **ML**, thereby supporting the complex formation in the synthesis.

AFM of ML

AFM provides very useful information on surface roughness, distributions and particle size of an either conductive or non-conductive matters [12]. In Figure 4 has shown the typical AFM images of **ML** measured on a glass slide. The topology images were obtained by scanning of an area of 51 × 51 μm^2. Notice that the **ML** has a rougher surface with arithmetic mean deviation thickness or lateral dimensions of lumps of complex is about 0.178 μm. Thus, the **ML** consists of large size sharp edges as is in good agreement with FESEM images shown in Figure 2.

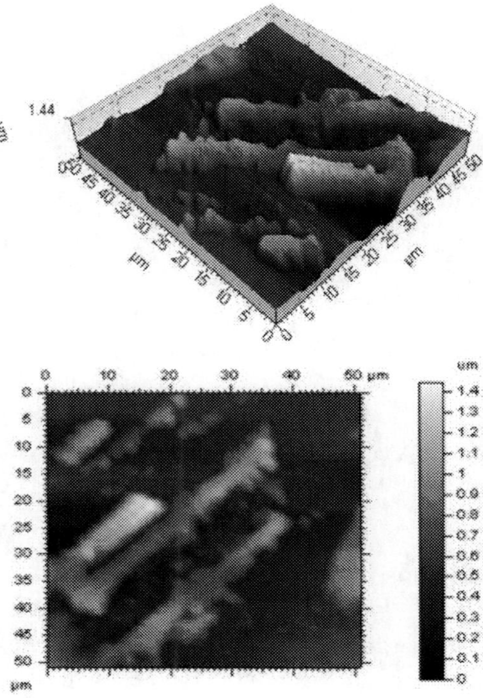

Figure 4. AFM images of the **ML** measured in the tapping mode. Height analysis versus distance shows a uniform thickness of about 0.178 μm.

Thermal Stability of ML

To establish the thermal stabilities as well as any solvent inclusion in the lattice of the **ML**, TGA and DTA analyses were carried out under nitrogen gas atmosphere. TGA of **ML** showed a single-step weight loss between 440–460 °C, which is attributed to the decomposition of the ligand, **L**. After decay, about 12.5% of the starting weight remains, and this residual material could correspond to the formation of copper oxides or metallic copper at high temperature [13]. It is worthy to mention that, no solvent or water molecule trapped in the crystal lattice as no weight loss until 440 °C has been noticed. This is in good agreement with EDS and IR data described above.

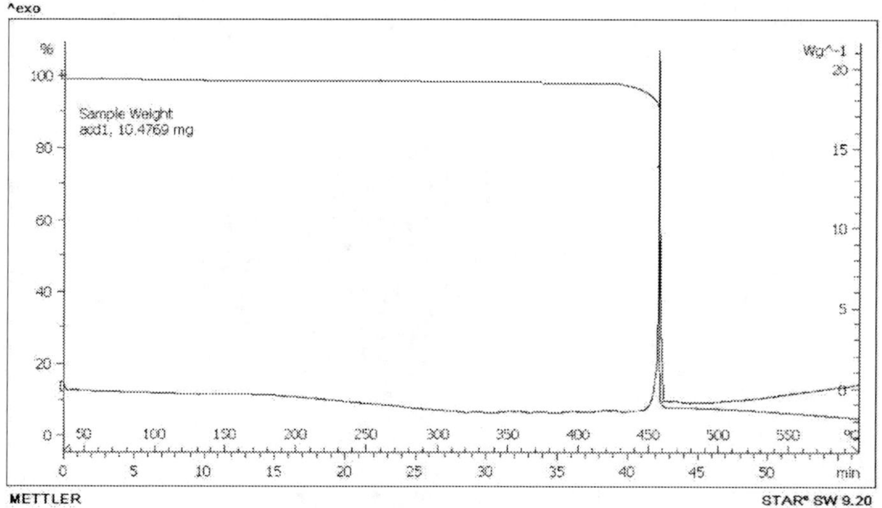

Figure 5. TGA and DTA curves of the **ML** recorded under nitrogen gas atmosphere at 10 °C/min.

Electrochemical Sensing of Caffeine by ML

Generally, a four coordinated first row transition metals have one or two additional coordination sites for more labile ligands [10]. As a result, an appropriate heterocyclic compound should be able to bind at these free coordination sites resulting in a change of the electronic environment of the complex and a change of the redox properties [13]. The change in redox potential should allow electrochemical detection of such organic compounds in solution [14], [15]. The dichloromethane solution of **ML** was voltammetrically titrated with various heterocyclic compounds, which are having detrimental effects on humans and the environment [16]. The cyclic voltammogram in the presence of caffeine displayed the anodic signal at + 0.71 V, significantly shifted from the signal at + 0.93 V that appeared in the absence of caffeine (Figure 6). Thus, the voltammetric response against the quantity of caffeine was investigated intricately to gain insight on the interactions between the caffeine and **ML** (Scheme 1) as taking caffeine above 400–500 mg at a time causes increased urination, muscle twitching, irritability, irregular or rapid heartbeat, anxiety, insomnia and so on [16].

Typical voltammetric characteristics of the sensor, i.e., **ML**, in response to caffeine are shown in Figure 6. One observed that, the initial oxidation potential of **ML** shifted gradually with the addition of caffeine, and attains a constant value at 0.21 M concentration of caffeine. The potential shift of the anodic peak was linearly correlated to the concentration of caffeine and saturated at + 0.71 V after the addition of 2.0 equivalences of caffeine revealing that the complex system has a 1:2 stoichiometry (Figure 6). Remarkably, a substantial shift in the potential was examined under the presence of micromolar amounts of caffeine thus suggesting the high sensitivity of the **ML** towards caffeine detection.

As mentioned already, the shift of oxidation potential may result from a binding of caffeine at the free axial sites of Cu(II) ions of **ML** [15]. The basicity of caffeine increased the electron density of Cu(II) ions, resulting in the negative shift of the anodic signal [17]. Also, caffeine sensing by **ML** was proved by EPR analysis due to paramagnetic nature of Cu(II) ions [10]. Figure 7 shows typical EPR profiles of dichloromethane solution of **ML** with and without having caffeine (2 mole). Notice that the g value (2.1465 ± 0.0006 for **ML** and 2.1646 ± 0.0001 for **ML** + caffeine) and profile shape varied significantly with the addition of caffeine. Of course, the g values observed are in good agreement with the values reported for other four- and six coordinated Cu(II) complexes [18, 19]. Thus, the electrochemical as well as EPR studies imply that the **ML** is a novel sensor for caffeine. A mechanistic understanding may help to design a series of new cost-effective electrochemical probes for detection of other alkaloids such as cocaine, retronecine, cytisine, nicotine and so on [16].

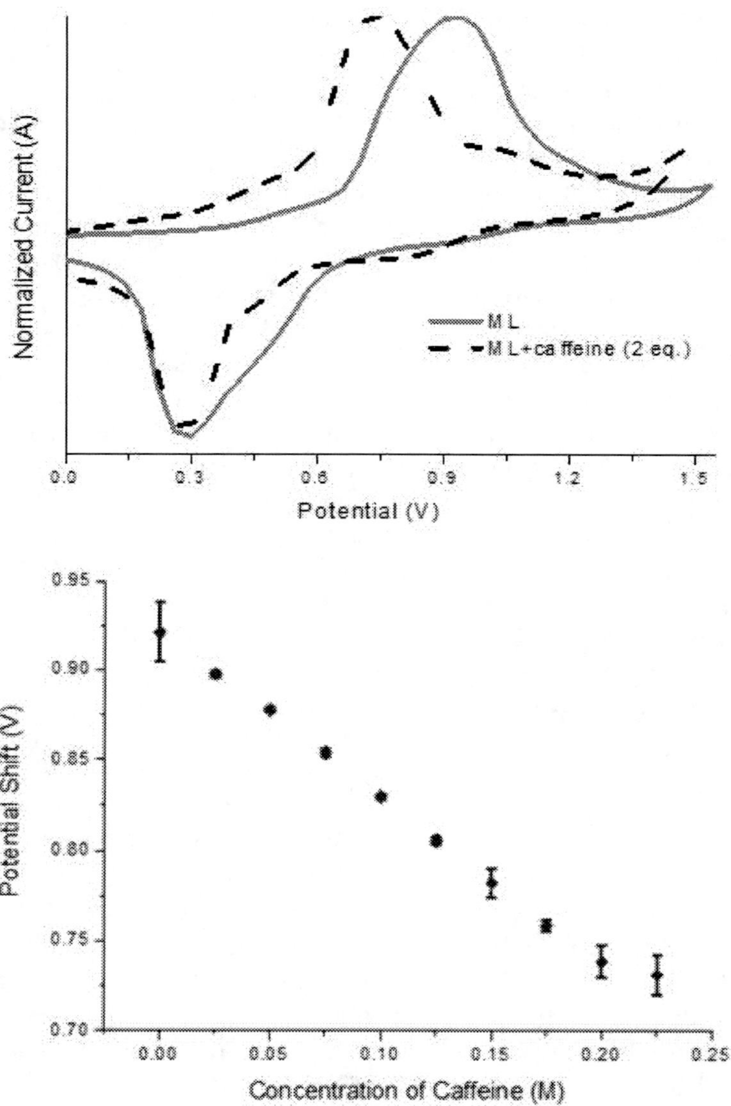

Figure 6. In left: the cyclic voltammograms of ML in dichloromethane solution in the absence and presence of 2 equiv of caffeine. In right: the plot between the potential shift of anodic signal and the concentration of caffeine for the dichloromethane solution of ML.

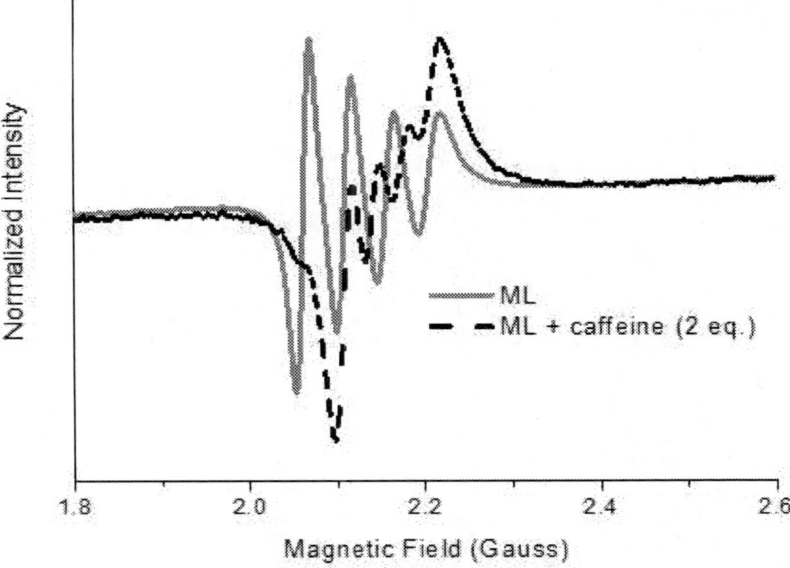

Figure 7. EPR profiles of **ML** in the absence (blue) and presence (red) of caffeine recorded in dichloromethane solution.

Conclusion

The conformationally flexible linker, 1,4-bis-(3-hexafluoroacetylacetonate)benzene, seems to be a versatile coordinating ligand that promptly react with Cu(II) ions yielding a coordinatively unsaturated network. The developed FMOF shows good solubility as well as electrochemical responses to caffeine a xanthine alkaloid at ambient conditions. Thus, we believe that the FMOFs based on flexible ligands surely will exhibit interesting functional behaviours in addition to the neat solubility and redox stability.

Acknowledgments

We gratefully acknowledge the financial support of the Department of Science & Technology (DST), India through INSPIRE Faculty Fellowship [IFA12-CH-62].

References

[1] MacGillivray, L. R. *Metal-Organic Frameworks: Design and Application*, John Wiley & Sons, Hoboken, New Jersey, 2010. https://doi.org/10.1002/9780470606858.

[2] Furuya, T. A. S. Kamlet, T. Ritter, Catalysis for fluorination and trifluoromethylation, *Nature,* 473 (2011) 470-477.

[3] Yang, C. X. Wang, M. A. Omary, Crystallographic observation of dynamic gas adsorption sites and thermal expansion in a breathable fluorous metal-organic framework. Angew. *Chem. Int. Ed.* 48 (2009) 2500-2505.

[4] Yang, C. U. Kaipa, Q. Z. Mather, X. Wang, V. Nesterov, A. F. Venero, M. A. Omary, Fluorous metal-organic frameworks with superior adsorption and hydrophobic properties toward oil spill cleanup and hydrocarbon storage, *J. Am. Chem. Soc.* 133 (2011) 18094-18097.

[5] Nijem, N. P. Canepa, U. Kaipa, K. Tan, K. Roodenko, S. Tekarli, J. Halbert, I. W. H. Oswald, R. K. Arvapally, C. Yang, T. Thonhauser, M. A. Omary, Y. J. Chabal, Water Cluster Confinement and Methane Adsorption in the Hydrophobic Cavities of a Fluorinated Metal–Organic Framework, *J. Am. Chem. Soc.* 135 (2013) 12615-12626.

[6] Pachfule, P. C. Dey, T. Panda, R. Banerjee, Synthesis and structural comparisons of five new fluorinated metal organic frameworks (F-MOFs), *Cryst. Eng. Comm.* 12 (2010) 1600-1609.

[7] Lin, Z.-J. J. Lu, M. Hong, R. Cao, Metal–organic frameworks based on flexible ligands (FL-MOFs): structures and applications, *Chem. Soc. Rev.* 2014, Doi: 10.1039/c3cs60483g.

[8] Luisi, B. S. V. Ch. Kravtsov, B. D. Moulton, An (8,3)-a 3D Coordination Network and Concomitant Three-Connected Supramolecular Isomers, *Cryst. Growth* Des. 6 (2006) 2207-2209.

[9] Field, L. M. P. M. Lahti, F. Palacio, A. Paduan-Filho, Manganese(II) and Copper(II) Hexafluoroacetylacetonate 1:1 Complexes with 5-(4-[N-tert-Butyl-N-aminoxyl] phenyl) pyrimidine: Regiochemical Parity Analysis for Exchange Behaviour of Complexes between Radicals and Paramagnetic Cations, *J. Am. Chem. Soc.* 125 (2003) 10110-10118.

[10] Sharma, R. K. *Text Book of Coordination Chemistry*, Discovery Publishing House, New Delhi, India, 2007.

[11] Pecharsky, V. K. P. Y. Zavalij, *Fundamentals of Powder Diffraction and Structural Characterization of Materials*, Springer, New York, USA, 2009.

[12] Haugstad, G. Atomic Force Microscopy: *Understanding Basic Modes and Advanced Applications*, John Wiley & Sons, Hoboken, New Jersey, 2012.

[13] Khedr, A. M. H. M. Marwani, Synthesis, Spectral, Thermal Analyses and Molecular Modeling of Bioactive Cu(II)-complexes with 1,3,4-thiadiazole Schiff Base Derivatives. Their Catalytic Effect on the Cathodic Reduction of Oxygen, *Int. J. Electrochem. Sci.* 7 (2012) 10074-10093.

[14] Heller, A. B. Feldman, Electrochemical Glucose Sensors and Their Applications in Diabetes Management, *Chem. Rev.* 108 (2008) 2482-2505.

[15] Korotcenkov, G. S. D. Han, J. R. Stetter, Review of Electrochemical Hydrogen Sensors, *Chem. Rev.* 109 (2009) 1402-1433.

[16] Joule, J. A. K. Mills, *Heterocyclic Chemistry*, John Wiley & Sons, Hoboken, New Jersey, 2013.

[17] Bard, A J. L R. Faulkner, Review: *Electrochemical Methods: Fundamentals and Applications,* John Wiley & Sons, Hoboken, New Jersey, 2001.

[18] Lee, Y.-M. Y.-K. Kim, H.-C. Jung, Y.-I. Kim, S.-N. Choi, Copper(II) Oxyanion Complexes Derived from Sparteine Copper(II) Dinitrate: Synthesis and Characterization of 4- and 5-coordinate Copper(II) Complexes, *Bull. Korean Chem. Soc.* 23 (2002) 404-412.

[19] Kozlevcar, B. P. Segedin, *Structural Analysis of a Series of Copper(II) Coordination Compounds and Correlation with their Magnetic Properties, Croatica Chemica Acta*, 81 (2008) 369-379.

Chapter 5

Iron(III) Nitrate As a Catalyst in Acid-Catalyzed Reactions

M. José da Silva*
Chemistry Department, Federal University of Vicosa, Viçosa, Minas Gerais, Brasil

Abstract

Metal salts were efficient catalysts in different reactions such as esterification of the terpene alcohols with acetic acid and glycerol acetalization with acetone. Herein, the recent developments obtained in these two reactions using commercial metal salts as the catalysts and renewable origin alcohols (i.e., glycerol and terpene alcohol) as the substrates were discussed. Esters of terpene alcohols and glycerol acetals are valuable ingredients for fragrance, agrochemicals, and pharmaceutical industries. In all the reactions, high conversions and selectivity toward the goal products were achieved. The different aspects involved in these reactions such as temperature, catalyst load, time, and stoichiometry of reactants were assessed. The actuation of the transition metal catalysts as Lewis's acids was discussed in detail.

Keywords: metal nitrate catalysts, terpene alcohols, glycerol, esterification, acetalization

* Corresponding Author's Email: silvamj2003@ufv.br.

In: Transition Metals - An Overview
Editor: Veronica J. Avila
ISBN: 979-8-88697-646-5
© 2023 Nova Science Publishers, Inc.

Introduction

Catalysis

The concept of catalysis was introduced over 180 years ago by the Swiss chemist Jöns Jacob Berzelius, as a chemical process to increase the reaction rate (Fechete et al., 2012). A catalyst is a substance that increases the rate of a reaction, offering an alternative route where the activation energy is lower than that of uncatalyzed reactions. Although the catalyst participates in key steps of the reaction, it is not consumed at the end of the process. The increase in reaction rate achieved in presence of a catalyst is often significant from a practical viewpoint, therefore, many chemical reactions are only kinetically viable if carried out in the presence of a catalyst.

Catalysis has great importance for various industries, consequently, most industrial chemical processes involve the use of catalysts in at least one stage of production. In this context, the majority of catalysts industrially used are metal compounds. Catalytic processes can be classified as homogeneous or heterogeneous. The homogeneous process is characterized by the presence of substrate and catalyst in the same phase; differently, in a heterogeneous process, the substrate and the catalyst are in different phases.

A homogeneous catalytic process ideal to be studied should satisfy the following requirements: all the catalyst molecules should be dispersed in the same phase; the process should allow an unambiguous investigation of chemical and spectroscopic characteristics; the reaction should proceed through a clear and well-defined kinetic pathway; in the catalytic cycle, the steps should be separately appreciated, allowing the adequate characterization of the intermediates; and, finally, the catalyst should be easily modified for special purposes. All these aspects contribute to the reaction to achieve high conversions. Moreover, when the active sites are well characterized, their adequate control favours an improvement in selectivity. Comparatively to the heterogeneous processes, the reactions in the homogeneous phase are carried out under milder conditions, due to the higher availability of the active sites. These reasons justify the academic and industrial interest in homogeneous catalytic processes.

Notwithstanding, most industrial processes have been performed using heterogeneous catalysis. In these processes, reactants and catalysts are in different phases. Typically, the reactant is in a gaseous or liquid state, while the catalyst is a solid material. These solid-catalyzed reactions take place on the surface of the catalyst. Normally, heterogeneous catalysts are prepared after the support of the active phase on a porous material with a high surface area. Their main advantage compared to homogeneous catalysts is that they can be easily recovered and reused, generating a lower amount of effluents and residues.

Esterification Reactions

Large-scale production of acetates through bioprocesses (i.e., extraction from plants or fermentation) is costly because it requires many purification steps (Claon et al., 1994). Therefore, the use of chemical catalytic processes such as metal-catalyzed esterification reactions become an attractive option (Silva and Oliveira, 2021). Esterification reactions of the carboxylic acids with alcohols are reversible processes which give besides the ester, water as a by-product. These processes usually require the presence of a catalyst, which can also be a Lewis or Brønsted acid (Hazan et al., 2015).

Esterification reactions are normally catalyzed by common Brønsted acids such as H2SO4, H3PO4, HF, and HCl (Leng et al., 2012). Alternatively, solid Brønsted acids such as Keggin heteropolyacids (i.e., H3PW12O40, HPAs) have been also successfully used in these laboratory-scale reactions (Da Silva and Coronel, 2018). Although HPAs are less corrosive than liquid mineral acids, they require neutralization steps at the end of the processes, resulting in effluents and residues (Mazumder et al., 2015; Vafaeezadeh et al., 2014). In addition, they are soluble catalysts, an aspect that hampers their recovery and reuse.

Therefore, the use of Lewis acid salts as an alternative to the Brønsted acid catalysts becomes potentially attractive for ester synthesis. Lewis acids are not very corrosive and do not require neutralization steps after the process, which minimizes the generation of wastes (Lopez et al., 2005). Among the Lewis acid catalysts, the tin(II) halides have received a highlight. They efficiently catalyzed the esterification reactions of naturally occurring substrates such as glycerol, fatty acids, and triglycerides (Gonçalves et al., 2011; Ferreira et al., 2013). Recently, tin(II) chloride was an active catalyst in the process of converting glycerol to ketal (Da Silva et al., 2015). Other Lewis acid metal salts have been also used as catalysts. Metal nitrates are potentially attractive because they are less corrosive than liquid Brønsted acids, are more stable than enzymes, and have a lower commercial cost.

In this work, we will address the selective esterification of renewable origin alcohols (e.g., glycerol and terpene alcohols) with acetic acid, which is a fundamental reaction on laboratory and commercial scales.

Ketalization Reactions

The reaction of glycerol with acetone is a versatile synthetic tool for using glycerol as a substrate in the biorefinery concept (Thomas et al., 2011; Li et al., 2017). Typically, acetalization of glycerol is a reaction carried out in a homogeneous phase using Brønsted acid catalysts as HCl or H2SO4. Strategies to produce solketal using sulfuric acid as a catalyst demonstrated to be very simple and reliable, mainly because provides a nearly 100% yield of solketal, due to the recycling of glycerol (Dmitriev et al., 2016). Although inexpensive, these liquid mineral acids are not recyclable, highly corrosive, and require steps of neutralization, which have a high environmental impact (Corma and Garcia, 2003). Although the use of Brønsted acids in a low concentration minimizes the corrosion, the steps of treatment and product purification still are required (Dmitriev et al., 2018).

Catalytic Runs

The reactions were carried out in a 25 mL three-neck glass reactor with sampling septum, in a thermostatic bath with magnetic stirring, prepared in a standard way, dissolving the substrate in the given solvent, and then adding the catalyst. Soon after, the flask was coupled to the reflux system with a bath at the appropriate temperature (Da Silva and Mosqueira, 2016).

The reactions were monitored for periods of 6 hours, collecting aliquots periodically and analyzing them by gas chromatography (gas chromatograph model Shimadzu GC-2010 plus), equipped with a flame ionization detector and a Carbowax 20M capillary column (30 m x 0.25 mm x 0.25 μm). The analysis conditions were: 80°C (1 min), a heating rate of 20°C/min; final temperature 250°C (1 min); injector temperature of 250°C; detector temperature of 250°C.

The identification of the products was made by GC-MS analysis comparing with standards and with results found in the literature. The mass spectrometer used was the Shimadzu MS-QP 2010 plus model equipment operating in electronic impact mode at 70 eV coupled to a Shimadzu GC 2010 chromatograph.

Results and Discussion

Effect of Lewis Acid Catalyst in the Esterification Reaction of β-Citronellol with HOAc

The choice of metallic cations to be evaluated was made according to their high Lewis acidity. The hydrated salts Al(NO3)3, Fe(NO3)3, Zn(NO3)2, Cu(NO3)2, Mn(NO3)2, Co(NO3)2, Ni(NO3)2 and LiNO3 were selected and tested as the catalysts. Although the LiNO3 salt has almost no Lewis acidity, it was also evaluated for comparison (Da Silva and Mosqueira, 2016).

Table 1. Conversion and selectivity of nitrate metal salt-catalyzed esterification of β-citronellol with acetic acid[a,b]

Exp	Catalyst	Conversion (%)	Selectivity (%)
1c	$Al(NO_3)_3$	27	50
2	$Fe(NO_3)_3$	38	70
3	$Zn(NO_3)_2$	24	54
	$Ni(NO_3)_2$	3	30
4	$Cu(NO_3)_2$	25	26
5	$Co(NO_3)_2$	3	26
6	$Mn(NO_3)_2$	8	25
7	$LiNO_3$	3	24

[a]Reaction conditions: β-citronelol (4.79 mmol), HOAc (19.2 mmol), catalyst (10 mol %), 333 K, 3 h, CH3CN.

[b]Conversion and selectivity are determined by GC analysis.

OH + OH Catalyst O O + H_2O

β-citronellol β-citronellyl acetate

Scheme 1. Esterification reaction of β-citronellol with acetic acid.

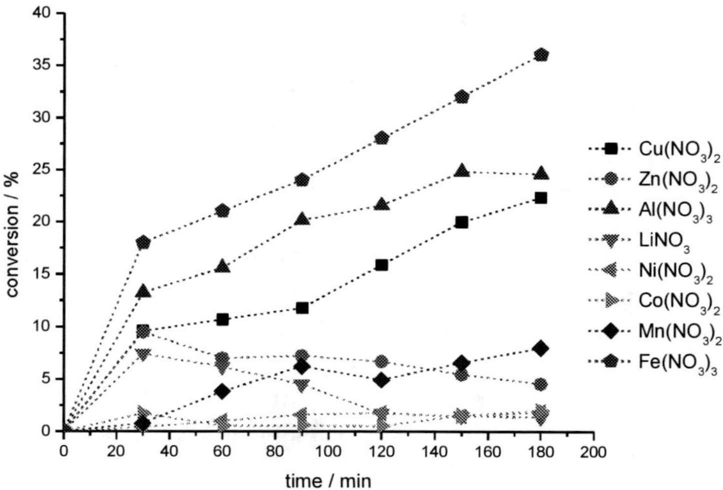

Figure 1. Kinetic curves of metal nitrate-catalyzed esterification reactions of β-citronellol with acetic acid (Da Silva and Mosqueira, 2016).

The β-citronellol is a monoterpene primary alcohol that has a trisubstituted double bond (Scheme 1). The presence of this double bond can compromise the ester selectivity, due to its susceptibility to undergo water addition.

The kinetic curves (Figure 1) showed that the Fe(NO3)3 catalyst was much more efficient than other metal nitrates. Remarkable, the iron(III) nitrate was the most active and selective catalyst, achieving the highest conversion and ester selectivity. Therefore, it was selected to evaluate the effect of the reaction variables. Initially, it was assessed the effect of the reactant stoichiometry (Table 2) (Da Silva and Mosqueira, 2016).

Besides the β-citronellyl acetate, oligomers were also obtained in all the reactions. The oligomerization is normally favoured in the presence of hydronium cations. The H3O+ cations were formed herein due to hydrolyse of Fe3+ cations of iron nitrate catalyst. Minority products were also detected, which are resulting from the water nucleophilic addition to the double bond of β-citronellol (Da Silva and Mosqueira, 2016).

The minimum molar ratio that provided the highest conversion was 1:12, therefore, it was selected to evaluate the effect of the catalyst load. Figure 3 shows the kinetic curves of reactions carried out with a variable load of the Fe(NO3)3 catalyst (Da Silva and Mosqueira, 2016).

Table 2. Effect of the molar ration on the conversion and selectivity of $Fe(NO_3)_3$-catalyzed esterification of β-citronellol with acetic acid[a,b]

			Selectivity (%)b		
Exp	**Molar ratio**	**Conversion (%)**	**β-citronellyl acetate**	**minority**	**oligomers**
1	1:1	28	94	1	6
2	1:4	37	70	15	15
3	1:8	48	71	6	22
4	1:10	51	73	9	18
5	1:12	57	67	14	18
6	1:16	57	64	15	22
7	1:20	57	59	19	21

[a]Reaction conditions: β-citronelol (4.79 mmol), HOAc (variable), catalyst (10 mol %), 333 K, 3 h, CH3CN.

[b]Conversion and selectivity were determined by GC analysis.

When a reaction is in equilibrium, an increase in the catalyst load does not affect the conversion. However, Figure 3 shows that within the interval of time studied, an increase in catalyst load increased the conversion of reactions. It is important to note that the focus herein was not to study the equilibrium reaction. The main goal was to verify what is the minimum concentration that allows the reaction to achieve the highest conversion.

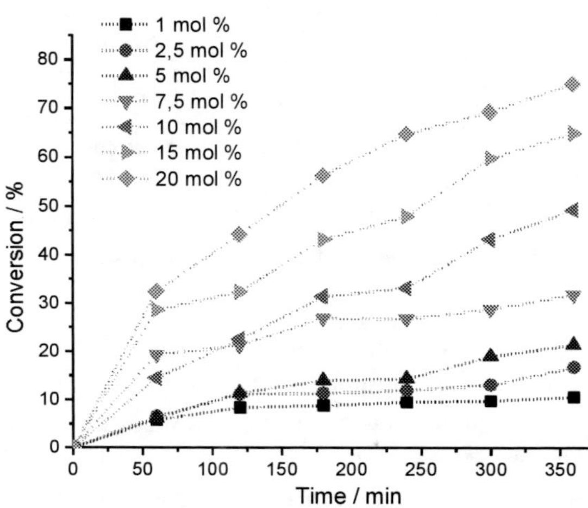

[a]Reaction conditions: β-citronellol (4.79 mmol), HOAc (38.32 mmol), catalyst load (variable), 333 K, 6 h, CH3CN.

Figure 3. Effect of $Fe(NO_3)_3$ catalyst load on kinetic curves of esterification of β-citronellol with acetic acid[a].

Although 20 mol % can be considered a large concentration, the low cost of the catalysts, as well as the possibility of their reuse, contribute to that it will be acceptable (Da Silva and Mosqueira, 2016).

Metal Nitrate Catalyzed Glycerol Condensation Reactions with Ketone

In this section, we will describe the results of the glycerol condensation reactions with acetone in the presence of the metal nitrates as the catalysts (Da Silva et al., 2020). The ketalization of glycerol with acetone is a reversible reaction that gave a mixture of two ketal glycerol isomers; the first one with a five-membered ring (i.e., solketal I, (2,2-6-dimethyl-1,3-dioxolan-4-yl)methanol)), and the other with a six-membered ring (i.e., solketal II, 2,2-dimethyl-1,3-dioxane-5-ol) (Scheme 4).

When the activity of metal nitrate was evaluated in the esterification reactions, $Fe(NO_3)_3$ and $Al(NO_3)_3$ were the most active and selective catalysts. Therefore, in the reaction of condensation of glycerol with acetone these two catalysts were evaluated. In Figure 4, a comparison of conversions achieved in the reactions with these two catalysts at different loads is depicted.

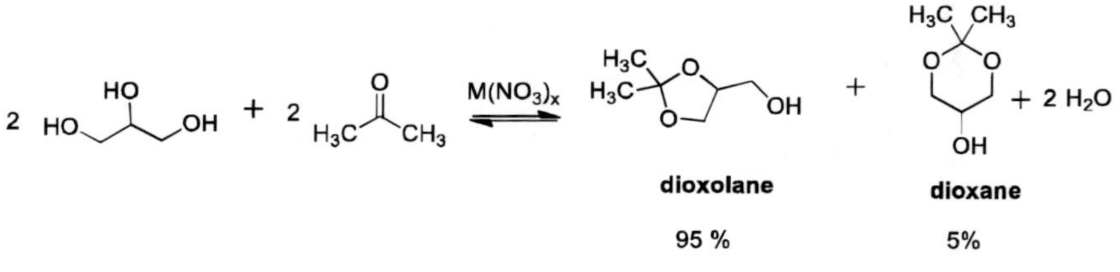

Scheme 2. Main products of metal nitrate-catalyzed glycerol ketalization reactions with ketone.

When used in low concentration, the $Fe(NO_3)_3$ catalyst was more active than Al(NO3)3. It can be assigned to its higher strength of Lewis acidity. In all the reactions, the proportion between the isomers of solketal was 95: 5 (dioxolane: dioxane).

The effect of the stoichiometry of the reactants was also evaluated. The reactions were carried out by varying the glycerol: acetone molar ratio from 1:1 to 1:25 (Figure 5). The maximum conversion was achieved when the proportion was 1:20. Moreover, the reaction selectivity was not impacted by the increase in the acetone load.

The activity of different metal nitrate was evaluated in this reaction and allowed to demonstrate that it is directly linked with the decrease in the pH triggered by solvolysis of metal nitrate catalyst (Table 3).

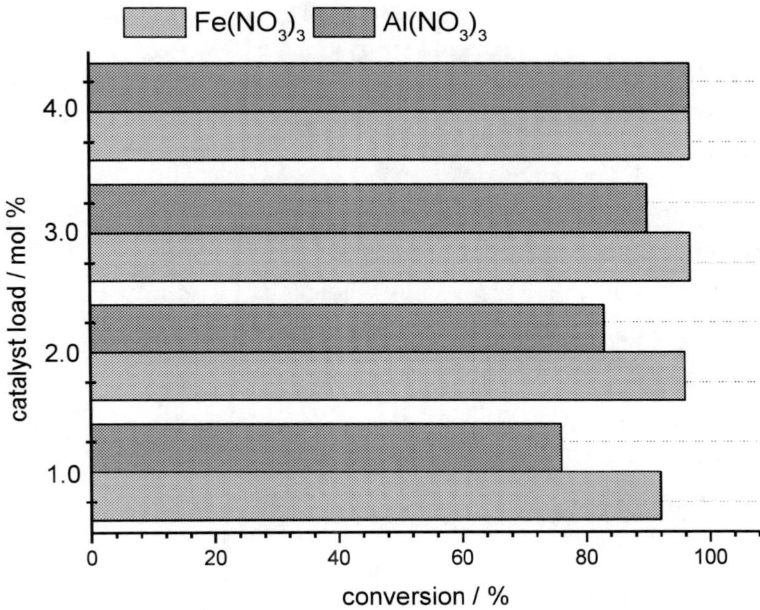

[a]Reaction conditions: glycerol (9.23 mmol), acetone (184.6 mmol), catalyst (0.3 mol %), temperature (298 K), reaction time (1 h), glycerol: acetone molar ratio (1:20) (DA SILVA et al., 2020).

Figure 4. Impact of catalyst load in the $Fe(NO_3)_3$ or $Al(NO_3)_3$-catalyzed glycerol ketalization reactions with acetone[a].

Table 3 shows clearly that the trends of activity of the catalysts (i.e., $Fe(NO_3)_3 \cdot 9\ H_2O > Al(NO_3)_3 \cdot 9\ H_2O > Cu(NO_3)_2 \cdot 3\ H_2O$) obeyed the decrease of pH provoked by the dissolution of metal salt in the acetone solution. Similarly to the observed in aqueous solution, the hydrolysis reaction can occur due to the presence of water molecules of hydration present in the catalyst, as well as the water molecules generated during the reaction.

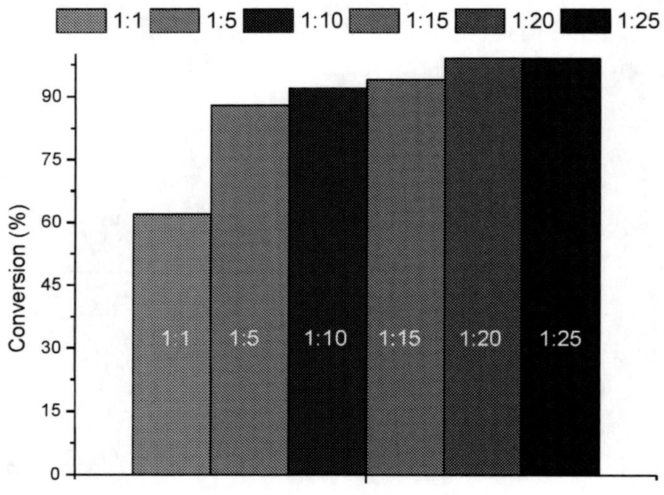

[a]Reaction conditions: glycerol (9.23 mmol), acetone (variable), catalyst (0.3 mol %), temperature (298 K), reaction time (1 h). (Da Silva et al., 2020).

Figure 5. Effect of the molar ratio of glycerol to acetone in the $Fe(NO_3)_3$-catalyzed glycerol ketalization reactions[a].

Table 3. Effect of catalyst on the pH and conversion of the condensation reactions of glycerol with acetone[a]

Run	Catalyst	pH in ketone solution	Conversion (%)	Selectivity[c] (%)	
		value		dioxolane	dioxane
1	none	6.0	0	0	0
2	Al(NO3)3·9 H2O	0.3	97	96	4
3	Co(NO3)2·6 H2O	2.2	0	0	0
4	Cu(NO3)2·3 H2O	0.9	32	86	15
5	Fe(NO3)3·9 H2O	-0.6	98	97	3
6	LiNO3	6.1	0	0	0
7	Mn(NO3)2·2 H2O	5.0	0	0	0
8	Ni(NO3)2·6 H2O	4.4	0	0	0
9	Zn(NO3)2·6 H2O	3.4	0	0	0

[a](Da Silva et al., 2020).

Conclusion

In this work, the efficiency of the metal nitrate catalysts was evaluated in two different types of acid-catalyzed reactions: esterification of terpene alcohols with acetic acid and glycerol acetalization with acetone. The focus was to evaluate how the nature of the metal cations present in the nitrate salts affects the acidity of these catalysts, and consequently, how impact their activity. Besides the Al^{3+} cations, transition metal cations such as Cu^{2+} and mainly Fe^{3+} demonstrated to be effective catalysts in these reactions. Remarkably, iron(III) nitrate efficiently catalyzed both reactions. These metal nitrate catalysts are potentially recyclable, inexpensive, and commercially affordable. Moreover, compared to the traditional liquid Brønsted acids, the metal nitrates are easier to handle and less corrosive, therefore, they are an attractive option to be used in different processes to convert renewable alcohols to more valuable products.

References

Claon, P. A., Akoh, C. C., *Enzyme Microbiology Technology,* 16, 835-838, 1994.

Corma, A., Garcia, H. *Chemical Reviews,* 103, 4307-4366, 2003.

Coronel, N. C., Silva, M. J., *Journal Cluster Of Science,* 29, 195-205, 2018.

Da Silva, M. J., Guimaraes, M. O., Julio, A. A., *Catalysis Letters,* 145, 769-776, 2015.

Da Silva, M. J., Mosqueira, D. A., *Catalysis And Science Technology,* 6, 3197-3207, 2016.

Da Silva, M. J., Pinheiro, P. F., Rodrigues, A. A., *Fuel,* 118164-118177, 2020.

Dmitriev, G. S., Terekhov, A V., Zanaveskin, L. N., Khadzhiev, S. N., Maksimov, A. L., *Russian Journal Applied Chemistry*, 89, 1619-1624, 2016.

Dmitriev, G. S., Zanaveskin, L. N., Terekhov, A V., Samoilov, V. O., Kozlovskii, I. A., Maksimov, A. L., *Russian Journal Applied Chemistry,* 91, 1478-1485, 2018.

Fechete, I., Wang, Y., Védrine, J. C. *Catalysis Today,* 189, 2-27, 2012.

Ferreira, A. B., Cardoso, A. L., Da Silva, M. J., *Catalysis Letters,* 143, 1240-1246. 2013.

Gonçalves, C. E., Laier, L. O., Da Silva, M. J., *Catalysis Letters,* 141, 1111-1117, 2011.

Hasan, Z., Yoon, J. W., Jhung, S. H., *Chemical Engineering Journal,* 278, 105-112, 2015.

Leng, Y., Jiang, P., Wang, J., *Catalysis Communications,* 25, 41-44, 2012.

Li, H., Yang, T., Riisager, A., Saravanamurugan S., Yang, S., *Chemcatchem* 9:1097-104, 2017.

Lopez, D. E., Goodwin, J. G., Bruce, D. A., Lotero, E., *Applied Catalysis A: General,* 295, 97-105, 2005.

Mazumder, N. A., Rano, R., Sarmah, G., *Journal Of Industrial And Engineering Chemistry,* 32, 211-217, 2015.

Silva, M. J., Oliveira, C. M., *New Journal Chemistry,* 45, 3683-3691, 2021.

Thomas, B, Ramu, V. G., Gopinath S, George J, Kurian M., Laurent, G, Sugunan S., *Applied Clay Science*, 53, 227-235, 2011.

Vafaeezadeh, M. and Hashemi, M. M., *Chemical Engineering Journal,* 250, 35-41, 2014.

Index

G

H

I

L

M

N

O

P

R

S

T

Y